AF849311

# Cancer Risks Associated With Elevated Levels of Drinking Water Arsenic Exposure

Prepared by:
**Floyd J. Frost**
Lovelace Respiratory Research Institute
2425 Ridgecrest Dr. SE, Albuquerque, NM 87108

Jointly sponsored by:
**Awwa Research Foundation**
6666 West Quincy Avenue, Denver, CO 80235-3098
and
**U.S. Environmental Protection Agency**
Washington, DC

Published by:

Awwa Research Foundation
Advancing the Science of Water®

IWA Publishing

## DISCLAIMER



**Published by IWA Publishing, Alliance House, 12 Caxton Street, London SW1H 0QS, UK**

Telephone: +44 (0) 20 7654 5500; Fax: +44 (0) 20 7654 5555; Email: publications@iwap.co.uk;
Web: **www.iwapublishing.com**

AwwaRF report number 91029F. Originally published by AwwaRF for its subscribers in 2004.
IWA Publishing version published 2005.



ISBN 1 84339 917 2

# CONTENTS

LIST OF TABLES ... vii

LIST OF FIGURES ... ix

FOREWORD ... xi

ACKNOWLEDGMENTS ... xiii

EXECUTIVE SUMMARY ... xv

CHAPTER 1 INTRODUCTION ... 1
Overview ... 1
Arsenic Exposure ... 1
Medicinal Uses of Arsenic ... 1
Adverse Health Effects of Arsenic ... 1
EPA Arsenic Maximum Contaminant Level (MCL) ... 8
Study Aims ... 9

CHAPTER 2 METHODS PART 1: UNDERLYING STUDY DESIGN ISSUES ... 11
Ecological Study Design ... 11
Multi-Level Models ... 12
Ascertainment of Cancer Outcomes ... 13
Accuracy of Death Certificate Data ... 14
Impact of Migration ... 15

CHAPTER 3 METHODS PART 2: SELECTING STUDY COHORT ... 17
Identification of Arsenic-Exposed Populations ... 17
Sources of Arsenic Occurrence Data ... 17
EPA Arsenic Occurrence and Exposure Database (AOED) ... 17
Additional State, County, and Utility Data ... 19
NRDC Database ... 19
Engel and Smith Data ... 19
Calculation of County Mean Arsenic Levels ... 19
Population Covered by Captured Arsenic Exposure Data ... 24
Person-Years of Exposure ... 25
Statistical Power ... 27
Counties In Final Study Cohort ... 27
Arsenic-Exposed Counties ... 27
Selection of Comparison Counties ... 28

CHAPTER 4 METHODS PART 3: STATISTICAL ANALYSIS ... 29
Analytical Models ... 29
Model Levels ... 30
Health Outcomes Data Used In Models ... 30
Cancer Mortality Data ... 30
Cancer Incidence Data ... 31
Standardized Mortality Ratios (SMR) ... 31

Standardized Incidence Ratios (SIR) .... 34
Explanatory Covariate Variables .... 36
Metropolitan Area .... 36
Socioeconomic Variables .... 36

CHAPTER 5 RESULTS AND DISCUSSION .... 39
Cancer Mortality Analysis .... 39
Bladder Cancer Mortality .... 39
Lung Cancer Mortality .... 43
Effect of Neighboring County Adjustment .... 43
Cancer Incidence .... 48

CHAPTER 6 SUMMARY AND CONCLUSIONS .... 61
Summary of Findings .... 61
Conclusions and Limitations .... 61
Implications of The Findings .... 62
Future Research .... 63

CHAPTER 7 RECOMMENDATIONS TO UTILITIES .... 65

REFERENCES .... 67

ABBREVIATIONS .... 72

# TABLES


3.1 State compliance monitoring data included in the EPA AOED .................................. 18

3.2 Number of wells with elevated arsenic .................................................................. 21

3.3 Mean drinking water arsenic concentrations by data source for U.S. counties with a mean arsenic concentration of 10 μg/L or greater .................................. 22

3.4 Counties with arsenic concentrations 10 μg/L or greater .......................................... 24

3.5 Person-years of exposure by decade for counties, with >10 μg/L arsenic exposure ..... 25

3.6 Person-years of exposure by decade, age and arsenic exposure group......................... 26

3.7 Sample size needed to detect a given relative risk (RR)* (alpha = 0.05, power 0.90, ratio unexposed to exposed = 4) ........................................................ 27

4.1 Bladder cancer mortality 1950-1999 within study cohort by sex and age.................... 30

4.2 Lung cancer mortality 1950-1999 within study cohort by sex and age........................ 31

4.3 Bladder cancer incidence 1973-1999 by sex .............................................................. 31

4.4 Lung cancer incidence 1973-1999 by sex.................................................................... 31

4.5 Bladder cancer standardized mortality ratio (SMR) by decade and drinking water arsenic level ....................................................................................................... 32

4.6 Lung cancer standardized mortality ratio (SMR) by decade and drinking water arsenic level ....................................................................................................... 33

4.7 Bladder and lung cancer standardized mortality ratios (SMR) 1950-1999 by arsenic level ....................................................................................................... 34

4.8 Bladder cancer SIR by decade and arsenic level ....................................................... 35

4.9 Lung cancer standardized incidence ratio (SIR) by decade and drinking water arsenic level ....................................................................................................... 35

4.10 Census variables used in statistical models ................................................................ 37

5.1 Bladder cancer mortality, males and females, all decades combined.......................... 40

5.2 Bladder cancer mortality, males and females age >50 years, all decades combined .... 41

5.3 Bladder cancer mortality in males and females by individual decades ........................ 42

5.4 Lung cancer mortality, males and females, all decades combined .............................. 44

5.5 Lung cancer mortality, males and females > 50 years, all decades combined ............. 45

5.6 Lung cancer mortality, males and females, by individual decades............................... 46

5.7 Bladder cancer mortality, county and neighboring area variation................................ 47

5.8 Lung cancer mortality, county and neighboring area variation .................................... 47

5.9 Bladder cancer incidence 1973-1999, males and females ............................................ 58

5.10 Lung cancer incidence 1973-1999, males and females ................................................ 59

# FIGURES


5.1 Bladder cancer SMRs in females, 1950-1999 .......... 49

5.2 Bladder cancer SMRs in males, 1950-1999.......... 50

5.3 Bladder cancer SMRs 1950-1999 in females age ≥ 50 years .......... 51

5.4 Adjusted bladder cancer SMRs 1950-1999 in females (observed and expected values increased by 1 prior to calculating SMR) .......... 52

5.5 Bladder cancer SMRs 1950-1999 in males age ≥ 50 years .......... 53

5.6 Lung cancer SMRs 1950-1999 in females.......... 54

5.7 Lung cancer SMRs 1950-1999 in males.......... 55

5.8 Lung cancer SMRs 1950-1999 in females age ≥ 50 years.......... 56

5.9 Lung cancer SMRs 1950-1999 in males age ≥ 50 years.......... 57

# FOREWORD

The Awwa Research Foundation is a nonprofit corporation that is dedicated to the implementation of a research effort to help utilities respond to regulatory requirements and traditional high-priority concerns of the industry. The research agenda is developed through a process of consultation with subscribers and drinking water professionals. Under the umbrella of a Strategic Research Plan, the Research advisory Council prioritizes the suggested projects based upon current and future needs, applicability, and past work; the recommendations are forwarded to the Board of Trustees for final selection. The foundation also sponsors research projects through the unsolicited proposal process; the Collaborative Research, Research Applications, and Tailored Collaboration programs; and various joint research efforts with organizations such as the U.S. Environmental Protection Agency, the U.S. Bureau of Reclamation, and the Association of California Water Agencies.

This publication is a result of one of these sponsored studies, and it is hoped that its findings will be applied in communities throughout the world. The following report serves not only as a means of communicating the results of the water industry's centralized research program but also as a tool to enlist the further support of the nonmember utilities and individuals.

Projects are managed closely from their inception to the final report by the foundation's staff and large cadre of volunteers who willingly contribute their time and expertise. The foundation serves a planning and management function and awards contracts to other institutions such as water utilities, universities, and engineering firms. The funding for this research effort comes primarily from the Subscription Program, through which water utilities subscribe to the research program and make an annual payment proportionate to the volume of water they deliver and consultants and manufactures subscribe based on their annual billings. The program offers a cost-effective and fair method for funding research in the public interest.

A broad spectrum of water supply issues is addressed by the foundation's research agenda: resources, treatment and operations, distribution and storage, water quality and analysis, toxicology, economics, and management. The ultimate purpose of the coordinated effort is to assist water suppliers to provide the highest possible quality of water economically and reliably.

The true benefits are realized when the results are implemented at the utility level. The foundation's trustees are pleased to offer this publication as a contribution toward that end.

This report summarizes a research study that assessed whether U.S. populations exposed to elevated levels of drinking water arsenic are at an increased risk of bladder and/or lung cancer. The U.S. Environmental Protection Agency lowered the drinking water maximum contaminant level (MCL) for arsenic in 2001 based on concerns that low dose drinking water exposures increase the risk of occurrence and death from bladder and lung cancers. This is the first national study to evaluate whether U.S. exposed populations have, in fact, suffered increased health risks from drinking water arsenic.

Walter J. Bishop
Chair, Board of Trustees
Awwa Research Foundation

James F. Manwaring, P.E.
Executive Director
Awwa Research Foundation

# ACKNOWLEDGMENTS

The author of this report is indebted to the following individuals for their cooperation and participation in this project:

| | |
|---|---|
| Joseph Chwirka, | CH2M Hill |
| Bruce Thomson, | Department of Engineering, University of New Mexico |
| Kristine Tollestrup, | Department of Family and Community Medicine, University of New Mexico |
| Malcolm Siegel, | Sandia National Laboratory |
| John Stomp, | City of Albuquerque |

In addition, Melissa Roberts of the Lovelace Respiratory Research Institute was responsible for the statistical modeling. Hans Petersen, Tim Muller, and Susan Paine of the Lovelace Respiratory Research Institute assisted with compilation and computation of drinking water arsenic levels in counties. Judith Hurley of the Lovelace Respiratory Research Institute provided technical assistance and prepared and revised study documents, reports, and papers. The author appreciates the advice of the Project Advisory Committee (PAC) members: Jack Colford, MD, PhD, Associate Professor of Epidemiology, University of California School of Public Health; Lynda Knobeloch, Wisconsin Department of Health and Family Services; Richard Nagel, Manager, Water Quality Central Basin, Metropolitan Water District, Carson, Calif.; and Paul White, U.S. Environmental Protection Agency. The author is also grateful for the able assistance of Linda Reekie, AwwaRF project officer.

# EXECUTIVE SUMMARY

## INTRODUCTION

In 2001 the U.S. Environmental Protection Agency (EPA) reduced the maximum contaminant level (MCL) for arsenic in drinking water from 50 µg/L to 10 µg/L, with the new MCL to take effect in 2006. This new MCL affects many U.S. community water systems located in areas of the country with high naturally-occurring arsenic in surface and ground water. Because of the expense of removing arsenic and the large number of systems affected, the costs of compliance with the new standard will place an economic burden on many small communities. EPA estimates that implementation of the revised arsenic standard will prevent between 6.9 and 33 bladder and lung cancer deaths each year. Recent epidemiological studies in Argentina and the United States, however, raise questions about the scientific justification for the revised arsenic MCL. In addition, although recent studies have failed to detect expected arsenic-related cancer risks, there are indications that EPA intends to further lower the arsenic MCL to below 10 µg/L. Several states, including California and New Jersey, have already begun the process of lowering their arsenic MCL to below 10 µg/L and California is considering implementing a standard below 1 µg/L (1 part per billion).

This study identified U.S. counties in which the mean drinking water arsenic level was 10 µg/L or greater during 1950-1999. We conducted a multi-level, hierarchical analysis of county standardized mortality ratios (SMRs) and standardized incidence ratios (SIRs).

## RESEARCH OBJECTIVES

The objective of this study was to examine whether lung and bladder cancer incidence or mortality rates are elevated in United States populations consuming drinking water that exceeds the new EPA MCL for arsenic of 10 µg/L.

## APPROACH

This study took place in two phases. In the first phase, we identified U.S. populations exposed to elevated levels of arsenic in drinking water. In the second phase, we evaluated the relationship between lung and bladder cancer mortality and incidence in those populations and drinking water arsenic exposure for 1950-1999.

To identify populations suitable for the evaluation of arsenic-related health effects based on exposure to arsenic in drinking water, we first evaluated existing arsenic occurrence data for U.S. drinking water systems. Primary data sources examined were the EPA Arsenic Occurrence and Exposure Database (AOED) and the National Resources Defense Council's (NRDC) arsenic database. We also examined additional published data on arsenic occurrence in U.S. counties. Because these various sources of data were incomplete and/or conflicting, we collected additional arsenic occurrence data directly from states, counties, and utilities. After comparing arsenic occurrence data across these multiple sources and determining the most accurate arsenic values for each county, we calculated mean drinking water arsenic levels for counties.

We next identified counties with average drinking water arsenic levels of $\geq$ 10 µg/L and >20 µg/L. This process resulted in the identification of 32 counties in 11 states in which the average arsenic concentration of at least 75% of public wells exceeded 10 µg/L. Of these 32 counties, 11 had an average arsenic concentration that exceeded 20 µg/L and 2 had an average arsenic concentration that exceeded 50 µg/L arsenic. The exposed counties were located in

Arizona, California, Colorado, Idaho, Illinois, North Dakota, New Mexico, Nevada, Oklahoma, Texas, and Utah. A total of 634 unexposed counties from these same states that met appropriate study criteria served as comparison counties in the analysis. The methods used for identifying arsenic-exposed counties and calculating mean arsenic levels are described in detail in this report and have been previously published (Frost et al. 2003).

We obtained lung and bladder cancer mortality data for all study counties and cancer incidence data for Utah and New Mexico counties. We also collected socioeconomic data for the study counties from the U.S. Bureau of the Census publications for use as covariates in some statistical models.

Using multi-level, hierarchical statistical models, we then examined whether there is evidence of excess bladder cancer mortality, bladder cancer incidence, lung cancer mortality, or lung cancer incidence in the populations exposed to drinking water with arsenic levels $\geq 10$ µg/L. We used SAS statistical software version 8.2 to aggregate data for analysis. For hierarchical modeling, we used MLwiN statistical software, version 2.0 (Centre for Multi-level Modeling, Institute of Education, London, England UK).

Where possible, analyses were performed using a continuous, normally distributed response variable. For the cancer mortality analyses the response variable was the age-standardized mortality ratio (SMR). For the cancer incidence analyses, the response variable was the age-standardized incidence ratio (SIR). In modeling uncommon occurrences, such as bladder cancer or lung cancer incidence or mortality within individual decades, and where low values of expected occurrences are a source of more variation in the calculated ratios, we modeled actual counts of occurrences using a Poisson response.

The mortality analysis also used multi-level models of county SMRs within neighboring areas. We employed three approaches for the mortality analyses. In the first approach, we combined all cancer deaths for all ages across the decades for which data were available, 1950-1999. The second approach was a sub-analysis limited to the population age 50 years and older. In the third approach, we combined cancer deaths for those decades for which comparable census variables were available, 1960-1999. This was, therefore, a repeated measures analysis of decades within county. This third analysis more precisely adjusted for the covariate effects of socioeconomic variables on mortality rates, since in the combined decades analysis, rankings of counties with respect to socioeconomic variables were assumed to remain constant across the five-decade time period. The socioeconomic variables for one decade, the 1980s, were used as a measure of a county's status with respect to other counties for all the decades.

For the cancer incidence analyses, individual year data for 1973-1999 were aggregated into three decades. As in the third mortality analysis approach, we conducted a repeated measures analysis of decades within county. These analyses were limited to study counties in New Mexico and Utah, the only study states with National Cancer Institute Surveillance and End Results (SEER) registries. No counties in these states had a mean drinking water arsenic level above 19 µg/L.

## CONCLUSIONS

Arsenic in drinking water at levels >10 µg/L was not associated with greater mortality from bladder or lung cancer, nor was a higher level of arsenic associated with greater incidence of bladder or lung cancer. There was considerable variation between counties in both lung and bladder cancer mortality, and county lung and bladder cancer mortality rates were strongly related to neighboring county lung and bladder cancer mortality rates. This suggests that adjustment for neighboring county cancer mortality rates controls for many of the unmeasured confounding factors. Higher mortality rates for bladder and lung cancer were observed in counties designated as metropolitan and, for males, counties with a high percentage of persons employed in manufacturing. Lower mortality rates were observed in counties with higher mean educational levels and counties with a larger mean household size. These same covariate relationships were not apparent in the cancer incidence models.

This study did not find evidence of increased risk for lung or bladder cancer mortality or incidence from exposure to arsenic in drinking water. The findings are consistent with other recent studies of the health effects of low dose arsenic exposure.

## RECOMMENDATIONS

The study presented in this report provides additional information useful for assessing arsenic-related health risks in U.S. populations and can help regulators better assess the need to further lower the arsenic MCL. Although this study has no direct or immediate implications for drinking water utilities, its findings can help guide the water industry in evaluating the scientific data on arsenic health effects in the United States. The information presented here may also be useful to customers concerned about the safety of their drinking water.

## FUTURE RESEARCH

In addition to cancer, cardiovascular disease has been linked to drinking water arsenic exposure. A prior study by Engel and Smith (1994) suggested that geographic areas with elevated drinking water arsenic have elevated rates of cardiovascular disease. It is important that the relationship between cardiovascular disease and drinking water arsenic be examined further. This could be accomplished using a study design similar to that used here. Multi-level hierarchical models is a highly appropriate method for determining if areas with elevated drinking water arsenic have elevated rates of cardiovascular disease, as it permits adjustment for rates of cardiovascular disease in neighboring counties. There is also a need to assess the health risks of drinking water arsenic exposure through the development and application of biomarkers of arsenic exposure.

# CHAPTER 1
# INTRODUCTION

## OVERVIEW

This study was funded by the Awwa Research Foundation (AwwaRF) and conducted in 2001-2004. The primary objective was to examine whether lung and bladder cancer incidence or mortality rates are elevated in United States populations consuming drinking water that exceeds the new U.S. Environmental Protection Agency (EPA) maximum contaminant level (MCL) for arsenic of 10 μg/L.

## ARSENIC EXPOSURE

Humans can be exposed to arsenic through air, food, and water, and a zero level exposure is impossible to attain. Arsenic is an abundant, naturally occurring element with an average concentration of 2-5 mg/kg in the earth's crust (Hindmarsh and McCurdy 1986). This semi-metal has been used for various industrial purposes since the Bronze Age, when it was found useful for hardening copper. Today, arsenic is used in the manufacturing of glass, ceramics, clothing, cosmetics, semiconductors and pesticides. Arsenic in wood preservatives (as chromated copper arsenate) accounts for 90% of all arsenic use in the United States. Arsenic occurs in ground water due to the weathering of rock and natural dissolution of minerals.

### Medicinal Uses of Arsenic

Arsenic has been used as a curative and preventive medicine for centuries, with peak usage occurring in the mid-to-late $19^{th}$ century when arsenic trioxide (an inorganic, trivalent form of arsenic) was an ingredient in popular medicinal preparations used to treat epilepsy, malaria, syphilis, asthma, chronic skin eruptions, nervousness, and stomach ailments. The use of arsenic as a cure for syphilis was pioneered in 1910 and it remained the treatment of choice until the introduction of penicillin in the 1940's. In 1967, Fowler's solution (containing 1% potassium arsenate) was found effective for treating severe asthma (Harter and Novitch 1967). Prior to the advent of radiation therapy, arsenic trioxide was the only treatment available for most leukemias (Neubauer 1947). Today, arsenic trioxide is an FDA-approved treatment for acute promyelocytic leukemia, and recently there has been renewed interest in the use of arsenic to treat other cancers, with several clinical trials suggesting arsenic trioxide has broad therapeutic potential for not only leukemia, but a number of other hematologic malignancies (Liu and Han 2003).

### Adverse Health Effects of Arsenic

Adverse health effects from arsenic exposure have been described in the medical literature since the $18^{th}$ century. In 1728, Henckel described dermal lesions associated with arsenic exposure (Harper and Miranda 1990). In 1829, Paris found that copper smelter workers exposed to arsenical fumes developed a cancerous disease of the scrotum. In 1888, patients treated for skin disorders with arsenic medications were observed to develop skin cancer (Hutchinson 1888). Respiratory health effects in British workers exposed to arsenic were observed in 1899 (Harper and Miranda 1990).

In the 1930s, several studies suggested that occupational contact with organic arsenicals, particularly sodium arsenite used in sheep-dip, could produce both skin and lung cancers

(Hutchinson 1888). Hill and Fanning (1948) showed an association between arsenic exposure and excess skin cancers (as well as lung cancers) in workers of British factories producing sodium arsenite. Arsenic keratoses and skin cancer were observed in Reichenstein, Salesia (Neubauer 1947) and Cordoba Province, Argentina (Borgono et al. 1977); both locales have ground water contaminated with high levels of arsenic.

***Studies in Taiwan Populations***

Arsenic-related health effects have been studied extensively in an area on the southwestern coast of Taiwan. The population is engaged in farming, fishing or salt production. The people in this area are very poor and subsist on a diet low in animal protein and fat and high in rice and sweet potatoes. Artesian well water was used for drinking water from the early 1900's, when wells were drilled to provide a stable source of drinking water, until about 1966, when another source of water was made available for most of the area. The arsenic content in wells that were tested during their use ranged from 0.01 to 1.82 μg/L; however, arsenic levels in many wells are unknown, since their use was discontinued before arsenic testing was conducted. During and after the time these wells were in use, a striking epidemic of Blackfoot disease (BFD) occurred. BFD is an uncommon peripheral vascular disorder that results in gangrene of the extremities; it is similar to diseases caused by ergot alkaloid poisoning. BFD was found to occur in conjunction with skin cancer, arsenical keratosis and hyperpigmentation, suggesting that the BFD may have been caused by arsenic exposure (Tseng et al. 1968). The first cases of BFD were reported in 1954 and by 1977, 1,455 cases were recorded. The prevalence of BFD in 1968 was 8.9 per 1000 residents. Both BFD and arsenicism were limited to people drinking artesian well water with a variable concentration of arsenic (0.1 to 1.81 μg/L) (Lu 1990). Interestingly, although BFD prevalence outside the southwestern coastal Taiwan area ranges from non-existent to uncommon, two other cases of BFD have been detected in Taiwan in an area served by surface water with a low arsenic content (less than 30 μg/L).

Chiang et al. (1993) reported that the incidence of bladder cancer in the BFD area for 1981-1985 was approximately 10-fold higher than for other areas of Taiwan (23.5/100,000 versus 2.3/100,000 for all of Taiwan). They suggested that the most likely causal agents were the content of arsenic and the high concentration of 'fluorescent substances' in the water. Guo et al. (1997) also reported elevated rates of bladder cancer in the BFD area, as well as cancer of the kidney, ureter and urethra. The bladder cancer excess was observed for adenocarcinoma but not squamous cell carcinoma. They found no evidence of elevated risk for renal cell carcinoma or nephroblastoma, and concluded that arsenic-associated cancer risks for these organs were cell-specific. Chiou and colleagues (2001), studying a northeastern Taiwanese population exposed to drinking water arsenic, found that the relative risk for urinary cancer and transitional cell carcinoma in that population compared to a national Taiwan population was statistically significantly elevated only at drinking water arsenic concentrations $\geq 100\ \mu g/L$.

Using cancer registries and death certificates, Guo and Tseng (2000) reported the relationship between bladder cancer incidence and arsenic exposure from drinking water for 243 townships and between bladder cancer mortality and arsenic exposure for 10 townships in southwestern Taiwan. Arsenic exposure was grouped into the following strata: 50-80 μg/L; 90-160 μg/L; 170-320 μg/L; 330-640 μg/L; and >640 μg/L. They found no change in bladder cancer incidence or mortality for the first four exposure strata but found an increased risk in males and females ($p<0.01$) at an exposure >640 μg/L.

Morales et al. (2000) re-analyzed early data from southwestern Taiwan collected by Chen et al. (1992). Using the Taiwanese population as a comparison group, they calculated the relative risks for lung and bladder cancer. This study was used as the basis for the EPA 2001 arsenic standard (NRC 2001). Lamm et al. (2003) reanalyzed the same data and found that all of the elevated cancer risks occurred among users of artesian well water and that the elevated risks only occurred at arsenic concentrations in excess of 400 μg/L. As mentioned above, Guo and Tseng (2000) used a different analysis to achieve essentially the same finding.

Many questions thus remain about the source of the BFD area skin and internal organ cancers. People drinking water from the deep artesian wells suffer a high rate of BFD, and this has not been observed in other populations consuming similar levels of drinking water arsenic. A possibly confounding factor is that the artesian water in southwestern Taiwan has a variety of interesting characteristics. The water is believed to come from surface water that passes through sediments rich in organic matter. The water contains inflammable gas, humic substances and arsenic. Lu (1990) argued that the humic substances are probably formed as an indirect result of elevated concentrations of arsenic in the ground. Bates et al (1992) argues that the humic substances in the water are unlikely to cause cancer, since the arsenic concentration is strongly related to the cancer risk. However, he does not address the possibility that the humic substances enhance the carcinogenicity of arsenic and other carcinogens. There appear to be significant disagreements among researchers over the cause of BFD, and recent research suggests these humic substances are biologically active and can induce changes at the cellular level (Cheng et al 2003). Additionally, a study in the United States did not find a positive relationship between high arsenic levels in drinking water and excess skin cancer (Morton et al. 1976), although Calabrese points out that the levels of arsenic in the Taiwan drinking water greatly exceeded those in the U.S. drinking water (Calabrese 1983).

### *Lung Cancer*

Elevated lung cancer mortality rates have been observed in copper smelter workers (Blot and Fraumeni 1975, Enterline and Marsh 1982, Higgins et al. 1981), southern Rhodesian gold miners (National Research Council 1977), German vineyard workers using arsenic-containing pesticides (Roth 1957, Braun 1958), and factory workers in plants producing arsenical insecticides (Ott, Holder and Gordon 1974). Milham, however, compared lung cancer rates for the 13 U.S. counties with copper smelters to rates for all other counties in the same state for the same time period and observed no excess in lung cancer mortality (Milham 1978).

Several studies have demonstrated a gradient of increased lung cancer mortality with higher occupational exposure to arsenic (Lee and Fraumeni 1969; Mabuchi, Lilienfeld and Snell 1979; Tokudome and Kuratsune 1976). Pinto and colleagues (Pinto et al 1977; Pinto, Henderson and Enterline 1978), studying workers at the ASARCO Ruston facility, showed that the strong correlation between total lifetime exposure to arsenic and excess lung cancer mortality could not be accounted for by smoking habits.

Pershagen and colleagues. (Pershagen, Elinder, and Bolander 1977) studied lung cancer mortality in a region near the Ronnskar smelter in northern Sweden, which began operation in 1928 and emitted 1-3 tons of arsenic per day from 1930-1960. The lung cancer death rate for the population residing within 15 kilometers (km) of the smelter was higher to compared the rate in the population residing more that 200 km from the smelter, but was not different from the national lung cancer death rate for Sweden. Several studies have shown that arsenic-induced lung cancer has a long latency period, ranging from 34 to 51 years for various exposure categories (Ott, Holder and Gordon 1974; Lee and Fraumeni 1969, Axelson et al. 1978).

Blot and Fraumeni (1975) examined lung cancer mortality in 36 U.S. counties with copper, lead, or zinc smelters or refineries. Lung cancer mortality for 1950-1969, corrected for demographic variables, was significantly higher among males (17%) and females (15%) residing in these counties. Milham (1983) discussed limitations of the Blot and Fraumeni study. Smelting counties were not separated from refining counties, and the publication provided no quantitative data on arsenic exposure levels. Since lung cancer mortality varies by a factor of 2 from state to state, he suggested it would be more appropriate to compare counties with and without a smelter within same state.

Newman et al. (1976) studied lung cancer cell types in two Montana copper mining and smelting counties. The study detected an increase in lung cancer incidence in both men and women in the towns of Butte and Anaconda. No increased lung cancer mortality was observed in the county as a whole.

Lyon and colleagues (Lyon, Fillmore and Klauber 1977) used a population-based cancer registry to conduct a case-control study of lung cancer near a Salt Lake City copper smelter. They compared the distribution of residential distances from the smelter and observed no association between cancer and distance from the smelter. This study had the advantage of using incident lung cancer cases rather than deaths. People diagnosed with lymphoma served as controls, which may have been problematic since another study (Tokudome and Karatsune 1976) suggested lymphoma might be associated with arsenic exposure.

Polissar and colleagues (Polissar, Severson and Lee 1978) studied lung cancer incidence near the same Ruston ASARCO smelter. The exposure level was categorized based on the distance of the residence from the smelter and the ratio of the sulfur dioxide concentration to the background concentration for each census tract. No excess lung cancer risk was found for people living closer to the smelter or in tracts with high sulfur dioxide levels.

Cordier and colleagues (Cordier, Theraiult, and Iturra 1983) studied mortality patterns near a copper smelter in Rouyn-Noranda, Quebec. The study found excess mortality in men from lung cancer, chronic respiratory diseases, and diseases of the digestive systems. In women, only an excess of chronic respiratory diseases was seen. Unfortunately, cigarette smoking history of decedents was not obtained.

Frost et al. (1987) studied female lung cancer deaths in 1935-1969 in Pierce County, Washington, where the ASARCO Ruston copper smelter is located. Female lung cancer deaths were studied because there is a low background rate of lung cancer in women. Three geographical areas defined the exposure groups. In none of the exposure groups did the observed number of lung cancer deaths exceed the expected number. A nested case-control analysis found no statistically significant differences for cases versus controls in either duration of residence in the area or distance from the smelter (56% of decedents had lived for 50 or more years in the county prior to death, suggesting that migration was unlikely to have significantly reduced the power of the study to detect an effect).

Using death certificates, Matanoski et al. (1981) studied cancer mortality in Baltimore residents who had lived near a chemical plant that produced calcium and lead arsenate, arsenic acid, Paris green, and sodium arsenite. In 1966-1974, an increase was seen in the incidence of lung cancer in men living in the census tract where the plant was located. No increased incidence was seen during earlier time periods for males or during any time period for females. After removing from the analysis all individuals residing in the census tract who also worked in the chemical plant, lung cancer death rates remained elevated. Average soil arsenic level was 63 ppm in the census tract.

Rom et al. (1982) conducted a case-control study of lung cancer near an El Paso, Texas copper smelter. For 1944 to 1973, 575 lung cancer cases and 1,490 breast and prostate cancer controls were identified from records of the Texas Cancer Coordinating Council. No association between risk of lung cancer and the distance of residence from the smelter was observed.

Similarly, Greaves et al. (1981) found no relationship between distance of residence from a smelter (10 copper smelters and 1 lead-zinc smelter) and lung cancer diagnosis or death (cases) versus prostate, colon, and breast cancer diagnosis or death (controls).

Brown and colleagues (Brown, Pottern, and Blot 1984) conducted a case-control study of lung cancer in the vicinity of a large nonferrous smelter in eastern Pennsylvania. They found a 60 % increase in lung cancer risk after adjusting for smoking and occupation. A 20 % increase was seen among residents of areas outside the town, but within 10 km of the smelter stacks. Soil sampling revealed elevated concentrations of inorganic arsenic, cadmium, and other metals in locations close to the smelter.

A study in a Belgian population with moderate exposure to arsenic via drinking water (≤50 μg/L) and/or zinc smelter emissions found no increase in diseases of the nervous system, liver and heart, and cancers (Buchet and Lison 1998). The area in closest proximity to the smelter (>2 km) had arsenic drinking water levels in the range of 20-50 μg/L and a yearly mean air concentration of 0.3 μg/m$^3$.

A study in Chile related low to moderate levels of arsenic in drinking water to an increased risk of lung cancer (Ferreccio et al. 2000). The study found that drinking water arsenic dramatically increased the risk of lung cancer in both smokers and non-smokers. For smokers, arsenic appeared to act synergistically with arsenic to increase lung cancer risk. The study found that arsenic exposures in the 60-89 μg/L range resulted in a fourfold increased odds ratio (confidence interval 1.8-9.6) for lung cancer risk. Among smokers exposed to 50 to 199 μg/L, the odds ratio for lung cancer was18.6 compared to 5.9 for non-smokers. If correct, these elevated risks should be detectable in Fallon, Nevada where the arsenic levels average about 100 μg/L. A problem with the Ferreccio et al study is the selection of controls. The area lacks both a cancer registry and good information on the population size for various towns and villages. The cases were identified from eight public hospitals between November 1994 and July 1996. The authors were concerned that if controls were selected from the same hospital as the cases, there would be over-matching of exposure (i.e., the cases and controls might have the same arsenic exposures). The exposures vary widely between areas served by the different hospitals. To avoid overmatching, the authors created a list of all patients admitted to any hospital in the study area. They selected several different control groups from that list but used a different method for each. They also added controls from another study. In addition, the authors had some difficulty interviewing controls from some areas. The authors note that the selection of the control group(s) is a weakness of the study, but they nevertheless conclude, "this study provides strong evidence that ingestion of inorganic arsenic is associated with human lung cancer".

***Bladder and Kidney Cancer***

In a Finnish study, 61 bladder cancer cases and 49 kidney cancer cases were identified from 1981-1995 cancer registry records and matched on age and sex to 275 controls (Kurttio et al. 1999). Water samples were obtained from the wells used by subjects during the years prior to cancer diagnosis. Overall, no relationship was observed between arsenic exposure and risk of cancer for either bladder or kidney cancers, but when the analysis was restricted to exposure 3-9 years before diagnosis, they observed a 2.44 relative risk for cancer at the highest arsenic exposure level (0.5-64 μg/L).

Bates et al. (2003) conducted a case-control study of bladder cancer in an area of Argentina affected by arsenic. An earlier ecological study in this area indicated an elevated risk of bladder cancer (Hopenhayn-Rich et al. 1996). The study enrolled 114 newly diagnosed cases of bladder cancer in 1996-1999 and 114 controls matched by age, sex and county of residence. Information was collected by interview concerning drinking water sources used in the 40 years prior to diagnosis, as well as information on prior residences and other risk factors. No association was found between arsenic exposure and bladder cancer risk, although there was some indication of increased risk in smokers for well water arsenic consumed in the decade 51-60 years prior to cancer diagnosis. Bates noted that inorganic arsenic may potentiate cancer risks from other exposures, such as smoking. The authors noted that the negative findings were unexpected given the prior work that found a dose-response relationship between drinking water arsenic and bladder cancer in an ecological study (Hopenhayn-Rich et al. 1996).

***U.S. Studies of Low Dose Arsenic Exposure in Drinking Water***

Several studies conducted outside the United States have shown that while exposure to arsenic in drinking water is associated with various cancers, these effects do not occur until arsenic levels are at least 100 μg/L and often several-fold higher (for a review, see Brown and Ross, 2002). Although studies have consistently linked arsenic exposure to adverse outcomes at high doses (e.g., ≥ 600 μg/L in drinking water), findings related to adverse outcomes from the lower dose exposures seen in U.S. populations are more difficult to interpret.

Engel and Smith (1994) sought to identify U.S. counties with higher arsenic levels in drinking water (population-weighted mean arsenic level ≥ 5 μg/L). They identified 30 such counties and examined mortality over a 17-year period (1968-1984). Their data show that mortality risk (SMR) for all cancers combined and for lung cancers was 1.0 for those with drinking water levels of 5-10 μg/L and significantly less than 1.0 for those with higher drinking water arsenic levels (10-20 μg/L). For the highest exposure category (20 μg/L or greater) the SMR was 0.8 for males (significantly less than 1.0, $p<0.05$) and 1.1 for females (95% C.I. 0.9,1.2)

A study by Bates and colleagues (Bates, Smith, and Cantor 1995) failed to find an association between arsenic levels in U.S. drinking water and bladder cancer. The drinking water concentrations of arsenic in this study averaged only 5 μg/L (the total range was 0.5-160 μg/L). While this case-control study suggested that smoking might increase the risk of arsenic-induced bladder cancer, this observation was not consistent with respect to latency period. Also, statistical significance was established by a 90% confidence interval. A more stringent confidence interval of 95% is conventionally used to show significance (Hennekens and Buring 1987). The study may also have suffered from application of multiple statistical tests, making interpretation difficult.

Lewis et al. (1999) conducted a large (N= 4,058) ecological study in which an arsenic-exposed population in Utah was examined for increased cancer and non-cancer deaths. Drinking water arsenic concentrations ranged from to 3.5-620 μg/L, and averaged approximately 100 μg/L. This study is particularly attractive because the study cohort consisted mainly of members of the Mormon Church, who would be expected to have relatively low exposures to tobacco and alcohol, substances which potentially confound rates of arsenic-related diseases (particularly bladder and lung cancer). This study was also relatively detailed in documenting exposure based on number of years exposed to arsenic water concentrations (exposure was measured in ppb-years). Mormon Church ward records were used to identify the area of residence and thus the water supply for individuals in the study. No relationship was found between exposure to arsenic-contaminated drinking water and bladder and lung cancer. In fact, of note is that standardized mortality ratios (SMRs) for lung cancer were significantly decreased in both the high and low exposure categories. A small but statistically significant increase in prostate cancer (SMR=1.45, CI 1.07- 1.91) was observed, but only at the middle exposure level and it therefore could not be used to confirm a dose-response relationship between arsenic exposure and prostrate cancer.

In a population-based case-control study, Karagas et al. (2001) examined the relationship between toenail arsenic concentrations and basal cell carcinoma (BCC) and squamous cell carcinoma (SCC). The study included 587 BCC and 284 SCC cases diagnosed in New Hampshire from 1993-1996. No statistically significant association between toenail arsenic content and either BCC or SCC was observed, even for the highest arsenic exposure category (0.35-0.81 μg/g toenail arsenic content).

Prompted by concern regarding a leukemia cluster in Churchill County, Nevada, Moore, Lu, and Smith (2002) investigated the relationship between childhood cancer incidence and arsenic exposure in drinking water from 1979-1989. Over 327,000 Nevada children were grouped into low, medium, and high exposure categories (>10 μg/L, 10-25 μg/L and 35-90 μg/L). No statistically significant association between arsenic and any type of childhood cancer was found in any of the exposure groups, nor was any specific association with leukemia observed. It should be noted that leukemia has not been associated with arsenic exposure even in Taiwan, where exposure to arsenic was considerably higher.

Steinmaus et al. (2003) conducted a case-control study in Nevada and California. Cases diagnosed between 1994 and 2000 (N =181) were identified from cancer registries. Controls (N=328) were identified in two ways: controls age 65 years and older were identified from Health Care Financing Administration records and controls younger than age 65 were identified using random digit dialing. Both cases and controls were asked questions to identify each lifetime residence location and source of drinking water. Arsenic levels for each drinking water source were obtained from routinely conducted analyses by health/environment departments. The advantage of this study is that arsenic exposure was based on both the arsenic concentration of the drink water and the amount of water consumed per day. The bladder cancer odds ratio for those consuming >80 μg/L arsenic per day was 0.94 (95% confidence interval [0.56,1.57]). The authors concluded that, "Overall, no clear association was identified between bladder cancer risk and the exposures found in our study. Interestingly, the overall risks were below those predicted using data from the highly exposed population in Taiwan."

### *Pesticide Arsenic Exposure*

Tollestrup and associates (Tollestrup, Daling, and Allard 1995) examined mortality in orchard workers exposed to a lead arsenate pesticide from 1890-1940. Exposure to arsenic in this population was considerable, with urine arsenic averaging 56–100 μg/L. Despite the elevated body-burden, no statistically significant associations were seen between exposure and all causes of death. Excess cancers of the lymphatic system, however, have been reported in workers producing arsenical pesticides (Blejer and Wagner 1976; National Research Council 1977; Ott, Holder and Gordon 1974).

### *Other Studies of Interest*

One U.S. study examined the effect of elevated arsenic exposure during childhood on subsequent diagnosis of cancer. High levels of arsenic contamination have been found in the area adjacent to the former ASARCO copper smelter in Ruston, Washington. Tollestrup et al (in press) followed a cohort of over 3,000 children aged 2 to 14 who lived near the ASARCO Ruston copper smelter between 1910 and 1932 and evaluated cause of death 30-80 years later.

As a surrogate measure of arsenic exposure, the investigators used the number of years lived (0-<1.0 year, 1.0-3.9 years, 4.0–9.9 years, and >10 years) within a one-mile radius of the smelter stack. Although prior studies (Polissar et al. 1990, Milham 1978) reported that urinary arsenic concentrations in children living near the smelter averaged 81 μg/L and roughly correlated with distance of residence from the smelter (suggesting the children had relatively high arsenic exposures), Tollestrup and colleagues found no evidence of increased bladder or lung cancer mortality, even in the three highest arsenic exposure categories.

In addition to cancers of the lung, bladder, kidney, and skin, high arsenic exposure has been associated with hemangiosarcoma, angiosarcoma, cancer of the large intestine, and nasal cancer (National Research Council 1977, Axelson et al. 1978) as well as with liver cancer, prostate cancer, neurological problems, gastrointestinal conditions, diabetes, and vascular changes (Abernathy, Thomas and Calderon 2003).

Experimental studies using laboratory animals have not been able to clarify the mechanisms of arsenical carcinogenesis. Most studies using laboratory animals have not observed an excess of cancers even when arsenic was administered at maximally tolerated dosages for long periods (National Research Council 1977, Axelson et al. 1978, ATSDR 2000).

### *Synergism Between Arsenic Exposure and Cigarette Smoking*

Recent studies have found some indication that chronic smokers who are also exposed to arsenic contaminated drinking water may have a slightly higher risk for lung or bladder cancer (Bates et al 2003, Steinmaus et al 2003). It should be noted that smokers, even in the absence of arsenic exposure, are at greatly elevated risks of developing these cancers. Although these recent studies suggest there may be a synergism between smoking and arsenic health effects, the NRC did not conclude from its review that the waterborne arsenic risks were affected by smoking (NRC 2001 and NRC 1999).

## EPA ARSENIC MAXIMUM CONTAMINANT LEVEL (MCL)

In 2001, the U.S. EPA adopted a new maximum contaminant level (MCL) for drinking water arsenic. During the 2001 NRC panel's deliberations, the author received requests from a member of the NRC panel for data on lung and bladder cancer incidence and mortality for U.S. arsenic-exposed populations. These data were not available at the time, however, as no studies had yet evaluated whether U.S. residents exposed to elevated drinking water arsenic levels have elevated cancer risks. In an earlier 1999 report, the NRC stated that "there is sufficient evidence from human epidemiological studies in Taiwan, Chile, and Argentina that chronic ingestion of inorganic arsenic causes bladder and lung cancer, as well as skin cancer" (NRC 1999). The 1999 NRC panel also noted important limitations of the scientific data, however, and recommended further research be done to characterize a possible exposure-response relationship at low exposures to arsenic in drinking water. The 1999 NRC report emphasized that, "With minor exception, epidemiological studies for cancer are based on populations exposed to arsenic concentrations in drinking water of at least several hundred μg/L. Few data address the degree of cancer risk at lower concentrations of ingested arsenic". The study reported here is the first rigorous effort to determine if arsenic-exposed U.S. populations have elevated risks of developing or dying from arsenic-related cancers.

## STUDY AIMS

This study examines whether United States populations consuming water containing arsenic at levels exceeding 10 μg/L have an increased the risk of lung or bladder cancer or death from lung or bladder cancer. The study aims were:

1. To identify U.S. populations exposed to elevated drinking water arsenic (>10 μg/L).
2. To determine whether observed standardized mortality ratios (SMRs) exceed expected SMRs for lung and bladder cancers in arsenic-exposed counties.
3. To determine whether observed standardized incidence ratios (SIRs) exceed expected SMRs for lung and bladder cancers in arsenic-exposed counties.

The first phase of the research involved careful identification of counties with populations exposed to elevated drinking water arsenic. The second phase involved conducting multi-level (hierarchical) analyses to relate lung and bladder cancer mortality and incidence to drinking water arsenic levels.

# CHAPTER 2
# METHODS PART 1: UNDERLYING STUDY DESIGN ISSUES

## ECOLOGICAL STUDY DESIGN

The study reported here is an ecological design. Ecological studies relate observations of exposures and outcomes, determined in the aggregate, for groups of people. Ecological studies differ from other epidemiological studies that relate exposures to outcomes of individual persons. For example, an ecological study may find that a population with a high prevalence of cigarette smoking also has a high rate of lung cancer, suggesting that smoking is causally related to the risk of lung cancer.

Particularly pertinent to this discussion are ecological studies that relate geographic variation in disease incidence or mortality to geographic variation in explanatory or exposure variables. Ecological studies have often been used to relate environmental variables such as drinking water characteristics to mortality rates. Ecological studies have also been used to relate differences in socio-economic characteristics with disease or death rates, often relating a social-deprivation index, such as county poverty level, to health outcomes.

For example, an ecological study by Enterline and Stewart (1956) involved analysis of United States county-level mortality data for 1949-1951. The study found marked geographical variation in death rates from heart disease and concluded that place of residence might be an important risk factor for cardiovascular disease mortality. Subsequently, Kobayashi (1957) found a strong geographic relationship between cardiovascular disease and water hardness. More than 70 ecological and other types of studies have examined this relationship and most were able to replicate Kobayashi's original finding. Together with other information, these studies suggest that water magnesium and, perhaps, calcium concentrations in drinking water may be protective factors for heart disease death (Marx and Neutra 1997).

The advantage of ecological studies is that data for groups are often readily available. Surveys of group characteristics are often available, and mortality data are routinely available. Therefore, ecological studies can be conducted in very large populations at a low cost. This design is especially useful for environmental health studies that require a large population to detect uncommon health effects. In some situations neither a case-control nor a cohort study is feasible.

Because of their low cost, ecological studies can be useful as an exploratory tool to generate hypotheses for future analytical studies. The associations or lack of associations that are observed in ecological studies, however, must be viewed with caution. Neither theoretical nor empirical analyses have offered consistent guidelines for the interpretation of ecological analyses. It should be remembered that the health, exposure, and demographic statistics in these studies characterize population groups, and serious errors can result when it is assumed that inferences from an ecological study pertain to the individuals within the group (Greenland and Robins 1994a, 1994b). This is often referred to as an ecological bias.

The ecological bias can arise when relationships between groups of individuals are applied to individuals. The ecologic bias occurs, in part, because the group is not homogeneous with respect to the exposure. The magnitude and direction of an association at the individual level can be very different than that observed at the group level. Since the group may not be homogeneous with respect to the exposure, the average exposures for the group may not accurately reflect individual exposures. Associations derived from ecological studies can overestimate, underestimate or even estimate a reverse relationship between the exposure and outcome than the true relationship. Therefore, caution is needed when interpreting findings of ecological studies.

The success of ecological studies that relate drinking water characteristics (such as water hardness), to health outcomes may depend on the exposure (i.e. drinking water characteristics) being relatively homogeneous. Although generally true today, it was even more the case prior to the widespread use of bottled water, when few sources of drinking water were available other than tap water. The characteristics of tap water available in an area were thus likely to be the characteristics of the drinking water consumed, serving to reduce one of the major limitations of ecological studies.

## MULTI-LEVEL MODELS

When other study designs are either not feasible or prohibitively expensive, one approach to improving the information from ecological studies is to replicate the studies in different geographic areas with different populations or different types of populations with similar exposures. For example, ecological studies have related well water mineral content to risks of death from heart disease. After nearly 70 ecological studies, a relatively consistent pattern of findings from different countries by different investigators suggest that ground water mineral content, most likely magnesium level, is protective for sudden cardiac death. It is possible that this knowledge would not have resulted from studies using other approaches.

In epidemiological research we find that many exposures and outcomes have a complex hierarchical structure. This is especially the case for geographical analyses commonly conducted for ecological epidemiological studies. Data for individuals are commonly aggregated at an administrative boundary level, such as the city or county, and then further aggregated at a higher level, such as state or region. In general, subject data outcomes are assumed to be independent of influences from other subject data and completely unrelated to other subjects' outcomes. This is the 'independent, and identically distributed' assumption that is the foundation of most statistical analysis. With disease mortality and incidence data gathered and aggregated by administrative boundaries, and by time periods, there is bound to be positive correlation between time periods for a given county or city, and also between adjacent or neighboring counties and cities. There is interest in adjusting for this bias when examining the data, and also in identifying spatial trends, i.e., to determine whether there is a tendency for disease occurrence to be more prevalent in one regional area than another. This is, in fact, what Enterline and Stewart (1956) noticed and is readily apparent when examining disease maps that show large areas with unusually low or high mortality rates from particular causes. Langford et al (1999) discuss this in depth, pointing out that when examining the possibility that covariates and hierarchical effects are influencing an outcome, there are three categories in which the outcome may be influenced:

a. within-area effects: in this study, county population characteristics
b. hierarchical effects: grouping of smaller areas into larger areas for accountability
c. neighborhood effects: regional effects by areas proximate to each other; neighboring areas may share social, economic or geographic characteristics

Geographical analyses are not without problems. One frequently cited problem is the difficulty of using standardized risk ratios from small area populations (Langford et al. 1999). Since the risk ratio is the result of dividing counts of observed events by the counts of expected events, risk ratios for small area populations often display extreme variation. Risk ratios may be inordinately high or low compared to large area populations. The same issue is relevant to risk ratios associated with rare diseases.

Assuming relative risks are distributed normally and counts are large enough to not have the small count problem discussed above, analytical analysis is fairly straightforward using a continuous response variable. Otherwise, models are best approximated using a Poisson response model (Lawson et al. 2003). A Poisson distribution has the unique characteristic that the variance is equal to the mean. The Poisson distribution should, in theory, describe the random occurrence of disease cases in a population. However, two papers have noted that there was considerably more geographic variation in disease occurrence than the Poisson distribution predicted (Pocock, Cook, and Beresford 1981; Breslow 1984). It is usually the case that the variance is larger than the mean. This additional variation is called extra-Poisson variation, or overdispersion. Underdispersion can also occur.

Neighboring areas often have similar disease occurrence or mortality rates. Therefore, some of the extra-Poisson variation in disease occurrence could be related to characteristics of the neighboring areas. Clayton and Kaldor (1987) developed an approach for separating the extra-Poisson variation into two components. They labeled the first component "heterogeneity", which is extra-Poisson variation that is unrelated to geographic location. This variation is a random effect that is independent and normally distributed. They called the second component "clustering". The clustering component is extra-Poisson variation related to the disease occurrence rates in neighboring areas. It is also independently distributed, with a mean equal to the clustering components of the neighboring areas. Neighbors are frequently defined as geographically adjacent areas, but may also be defined by distance.

In addition, we have an explanatory component that may be studied for differences in effect between areas (random effect) or studied for uniform effect between areas (fixed effect). This explanatory component may relate to socio-class or occupational characteristics of residents. An example of an explanatory component is an index of urbanization that is related to the risk of disease occurrence or death. The introduction of the clustering component allows the calculation of a relative risk of disease, controlling for location. Location is considered to be a possible confounder and adjustment for location allows for more precise estimates of the effects. For example, smoking rates may vary geographically and by adjusting for rates of bladder or lung cancer in neighboring areas, the potentially confounding effects of smoking are controlled.

In the study reported here, all explanatory components have been modeled as fixed effects. The random part of the model pertains to the intercept value in the model, or initial value before adjusting for explanatory components. Drinking water arsenic level is, therefore, a fixed effect of interest. The effects of location on disease occurrence (cancer) in this study are likely to be related to differences in the prevalence of prior cigarette smoking, industrial exposures, and other causal factors that vary geographically. Our primary aim was to accurately estimate the effects of drinking water arsenic exposure after adjusting for other factors related to disease risk that also vary geographically. Comparing populations exposed to higher levels of arsenic with unexposed populations in the presence of the cluster term effectively allows us to estimate the effect of drinking water arsenic exposure among populations with similar exposures to other risk factors

## ASCERTAINMENT OF CANCER OUTCOMES

Epidemiological studies assume that the outcomes are completely ascertained. Complete ascertainment of illnesses or deaths from a disease is very difficult and there are only a few diseases for which outcomes can be completely ascertained. Although death registration in the United States has been relatively complete since the mid-1930's, the assigned cause of death can be inaccurate. To improve studies of cancer, the National Cancer Institute (NCI) established cancer registries in selected parts of the United States to provide complete and accurate estimates of cancer incidence for these areas. These registries serve as a basis for conducting etiological, survival and treatment studies (Surveillance, Epidemiology and End Results [SEER] registries). The SEER registries were begun in the early 1970's and considerable effort has been made to ensure complete ascertainment and to document the completeness of that ascertainment. More recently, the Centers for Disease Control and Prevention (CDC) has contracted with state health departments to establish cancer registries. These state registries have not had the experience of the NCI-sponsored registries and have only limited documentation of completeness of case-ascertainment. Also, since different states have begun collection of cancer incidence data at different times, the duration of available data collection is generally less than five years per state.

We focused most of our analysis on cancer mortality rather than cancer incidence because of the availability of longitudinal data. Only two states with elevated drinking water arsenic have SEER registries (New Mexico and Utah), with enough years of data to support a longitudinal study. Within these two states, only four counties have elevated drinking water arsenic levels. Cancer incidence data from non-SEER state registries are available for only a few years. Therefore, rather than limit our study to New Mexico and Utah SEER cancer incidence data, we focused primarily on cancer mortality in all exposed states. This allowed us to examine cancer outcomes over a larger sample of exposed populations.

To assess the accuracy of the cause of death in mortality data (obtained from death certificates), we examined the cause of death for lung and bladder cancer cases in the New Mexico and Utah SEER registries. This comparison permitted an assessment of the accuracy of cause of death on the death certificate and indicated the limitations in using mortality data to study the relationship between drinking water arsenic exposure and lung and bladder cancer mortality. SEER data for 1973-1999 include 22,0378 deaths in lung cancer cases. The death certificates of 17,049 (76%) of these SEER cases were also coded as death due to lung cancer on the death certificate. Among the SEER bladder cancer cases were 5,518 deaths, but only 29% of these deaths were coded as being due to bladder cancer on the death certificate. This lower bladder cancer death rate probably occurs because bladder cancer is less commonly a fatal disease than is lung cancer.

### Accuracy of Death Certificate Data

Accuracy and reliability of death certificate data has been a topic of debate over the years. Existing studies of this issue are difficult to compare due to differences in sample size, study design, and focus (Rosenberg 1989). Diagnoses of cancers have tended to be more accurate than other diagnoses (Rosenberg 1989). An early study validated death certificate information by examining medical information on 1,837 deaths in Pennsylvania (Moriyama et al. 1958). Additional information was also collected from the certifying physician. The investigators found that in only 14% of lung cancer deaths was cause of death unclear or without apparent justification. Even earlier, Dorn and Horn (1941) found that 85% of respiratory cancer deaths were properly coded to respiratory cancer on the death certificate.

Of particular relevance to this study was an analysis of 48,826 primary cancers by Percy, Stanek, and Gloeckler (1981). The study found a high rate of agreement between lung and bladder cancer hospital diagnoses and underlying cause of death codes on death certificates. Bladder cancer deaths had a detection rate of 91.1% and a confirmation rate of 93.6%. Trachea, lung, and bronchus cancer deaths had a detection rate of 95.0% and a confirmation rate of 93.9%. Detection rate was defined as the percentage of cases in which the number of persons with a hospital diagnosis of death due to a specific site of cancer agreed with death certificate data. Confirmation rate was the percentage of cases with a specific site of cancer on the death certificate confirmed by a hospital diagnosis.

**Impact of Migration**

A potential problem with ecological studies of arsenic-associated risks is that residents may move away from an arsenic-exposed area before either their cancer is diagnosed or their death occurs. In that case, the arsenic-related cancer or death occurs elsewhere and the cancer will not be counted as an observed case or cancer death in the exposed population. This reduces the power of studies to link the cancer case or death to the exposed population. For diseases with a long latency period, such as bladder and lung cancers, this underestimate of risk in high arsenic areas could be significant (Polissar 1980, Kliewer 1992). A related problem is that the latency of arsenic-related health effects is unknown. Studies have found latency periods between the first exposure to arsenic-contaminated drinking water and the development of skin lesions ranging from 5 to 23 years (Haque et al. 2003). As discussed in Chapter 1, some researchers suggest that the latency may be greater than 50 years. Little information is available for latency periods of other arsenic-related health outcomes or for the relationship between dose and latency.

Certain measures of migration, however, can overestimate the effect of migration, reducing the power of studies to detect environmental risks. For example, if the average migration rate for a population is used to estimate migration for elderly people, this likely overestimates the migration effect on study power. Lung cancer generally occurs after age 50 (mean age of occurrence is 71 years) and bladder cancer after age 60 (mean age of occurrence is 77 years). Migration rates generally decline with increasing age. Therefore, the migration rates for the 15 to 25 years prior to cancer occurrence or death will be lower than the migration rates for a 15 to 25 year period at a younger age. In a study in we conducted in Tacoma, Washington (Frost et al. 1987), we found that people resided in the county more than 30 years prior to their death. This generally agrees with a U.S. Bureau of the Census study (Schacter 2002) that found the mean duration of residence of people over 65 years of age was 18.7 years. A longer duration would be expected for residence in the same city. For a one-year period (1998-1999), the Bureau of Census found that over 55% of people age 65 years and older who changed residences moved to a residence in the same county. In addition, the effect of return migration is seldom considered. A study by Long (1988) categorized migrations as primary (those leaving the area of their birth for the first time), secondary (leaving an area that was not their area of birth) and return (returning to their area of birth). Between 1975 and 1980, 42% of migrants were primary, 38.8% were secondary, and 19.2% were return migration. Therefore, almost one fifth of migrants returned to the area of a prior residence. If migration were random, only 2% would have returned to their state of birth, according to Long. A large fraction of the return migration is the elderly returning to their state of birth. Alternatively, if the latency between completion of exposure and cancer occurrence is greater than 50 years, then migration could be important. However, if 20 years of exposure is required to increase the cancer risk and if an additional 50 years of latency is required before the onset of cancer, then arsenic-related cancers would not occur before age 70.

# CHAPTER 3
# METHODS PART 2: SELECTING STUDY COHORT

## IDENTIFICATION OF ARSENIC-EXPOSED POPULATIONS

To identify arsenic-exposed populations suitable for epidemiological studies of arsenic health effects, we examined drinking water arsenic occurrence data from several sources, compared the data between sources, and conducted a detailed review of systems identified as having arsenic-contaminated drinking water sources (Frost et al. 2003). The following summarizes the methods and findings of this previously published work.

## SOURCES OF ARSENIC OCCURRENCE DATA

To identify counties with elevated arsenic levels, we used three primary sources: the EPA Arsenic Occurrence and Exposure Database (AOED), data we obtained directly from states, and data we obtained directly from utilities and counties. We also examined, but did not use in our estimates of arsenic exposure, the NRDC database and data included in a publication by Engel and Smith (Engel and Smith 1994).

### EPA Arsenic Occurrence and Exposure Database (AOED)

To estimate the national occurrence of arsenic in drinking water, the EPA developed the Arsenic Occurrence and Exposure Database (AOED) (EPA 2000). The AOED is based on information from the EPA Safe Drinking Water Information System (SDWIS) and from state compliance monitoring data sets.

The list of water utilities in the AOED is derived from the SDWIS. This SDWIS is an inventory of water systems by state; it contains no data on levels of contaminants in those systems, however, it does list violations of maximum contaminant levels (MCLs). The SDWIS contains each utility's name, address, federal identification number, source water type, ownership category, population served, and regulation classification (system type). Although few large water systems are missing from the SDWIS, some small, privately owned drinking water systems that are not known to state regulatory agencies are not included. Systems that use both ground and surface water are classified as a surface source, since surface water has more stringent treatment requirements concerning removal or inactivation of pathogens and protection from disinfection by-produces. The population served by a given system is based on the number of retail customers and does not include the number of people served by other water systems that purchase water from a the utility. The population served is usually an estimate since neither the state nor the utility knows how many people reside in the residences served by the utility. The utility usually knows the number of residential and business connections.

Arsenic occurrence data in the AOED is derived from state compliance monitoring data. Each state maintains some form of a drinking water compliance-monitoring database. In its most simple form this database contains information on compliance with all current MCLs for regulated pollutants, including whether or not the water utility is in compliance with the previous arsenic MCL of 50 μg/L. Larger utilities are required to test more frequently than smaller utilities. Reported compliance with the MCL implies that analytical procedures with a method detection level (MDL) of less than the MCL were utilized. In some states the compliance-monitoring database also contains the actual concentration of the pollutant detected and, in fewer states, it identified the analytical method and detection levels.

Arsenic occurrence data from state compliance monitoring databases were provided to EPA by 32 states. The remaining states were not able to comply. Data from 7 of the 32 submitting states were not considered suitable for inclusion in the AOED because: 1) arsenic concentrations were truncated at a relatively high arsenic level, such as 20-50 µg/L, 2) arsenic concentrations were not included, 3) system identification numbers were not included, or 4) data were submitted on paper rather than computer-readable medium. Data from the remaining 25 states were included. Some states submitted multiple data sets. In that event, only the most recent data set was included.

The AOED contains data from most regions of the United States; however, relatively few states in New England, the Mid-Atlantic, and the Southeast are included. The states and the percentage of the state's ground water systems that are included in the database are shown in Table 3.1.

**Table 3.1**
**State compliance monitoring data included in the EPA AOED**

| State | No. of community water systems using ground water | No. of community water systems using surface water | Years included | Percentage of state ground water systems included |
|---|---|---|---|---|
| Alaska | 326 | 109 | 1991-97 | 96.7% |
| Alabama | 263 | 68 | 1985-00 | 97.4% |
| Arkansas | 371 | 76 | 1996-98 | 99.5% |
| Arizona | 668 | 46 | 1988-98 | 92.6% |
| California | 1369 | 222 | 1981-00 | 51.6% |
| Illinois | 1082 | 103 | 1993-00 | 95.8% |
| Indiana | 648 | 51 | 1996-99 | 81.6% |
| Kansas | 506 | 101 | 1992-97 | 98.1% |
| Kentucky | 88 | 150 | --- | 87.1% |
| Maine | 109 | 29 | 1991-94 | 34.6% |
| Michigan | 644 | 33 | 1993-97 | 56.4% |
| Minnesota | 863 | 23 | 1992-97 | 95.7% |
| Missouri | 773 | 89 | 1995-97 | 70.5% |
| Montana | 484 | 47 | 1980-92 | 89.0% |
| North Carolina | 1735 | 169 | 1980-00 | 93.9% |
| North Dakota | 197 | 19 | 1993-95 | 97.5% |
| New Hampshire | 504 | 37 | 1990-94 | 81.6% |
| New Jersey | 438 | 29 | 1993-97 | 88.7% |
| New Mexico | 573 | 29 | 1983-00 | 98.5% |
| Nevada | 221 | 31 | 1991-97 | 88.8% |
| Ohio | 875 | 139 | 1981-94 | 86.1% |
| Oklahoma | 446 | 210 | 1995-98 | 97.4% |
| Oregon | 583 | 134 | 1990-98 | 87.7% |
| Texas | 3105 | 326 | 1994-99 | 90.8% |
| Utah | 327 | 38 | 1980-99 | 97.9% |

Source: EPA (Environmental Protection Agency). 2000. *Arsenic Occurrence in Public Drinking Water Supplies*. EPA-815-R-00-023. Office of Ground Water and Drinking Water, Environmental Protection Agency. Washington D.C.: U.S. Environmental Protection Agency.

### Additional State, County, and Utility Data

To validate and expand on the AOED data, we obtained new raw data from selected state drinking water agencies, counties, and utilities. In addition to the states included in the AOED, we obtained data from some states not included in the AOED or NRDC databases: Alaska, Michigan, Nevada, Idaho, Oklahoma, Colorado, California, and Texas. We also obtained data from Pierce County, Washington; Union County, North Carolina; Ramsey County, North Dakota; and Green, Piatt and Gallatin counties, Illinois.

### NRDC Database

The NRDC also has a database on arsenic occurrence (Mushak 2000). The NRDC database started with arsenic concentrations for individual sources. These data were originally obtained from the EPA and may have been updated from information supplied directly by utilities. Perhaps because the NRDC database was assembled by volunteers, little documentation is available on how the database was assembled, what quality assurance tests were used in assembling or evaluating the data or whether the original data obtained from the EPA were subsequently updated. We used this data only for comparative purposes against other data sources, and did not use it in determining our estimates of county arsenic exposures.

### Engel and Smith Data

One published study provided arsenic exposure estimates for 30 U.S. counties (Engel and Smith 1994). This study found an association between arsenic exposure and the risk of cardiovascular death. According to the authors, the arsenic exposure data were obtained by requesting drinking water arsenic data for public water systems from all 50 states, the District of Columbia, and Puerto Rico. The mean arsenic level per county was computed by weighting the mean arsenic level per water system by the size of the population served by the system. In several counties with large populations, however, we found poor agreement between this study's reported county arsenic levels with those that we calculated from data in the AOED and NRDC database. In an attempt to resolve these differences, we contacted local public health authorities. We were unable to find evidence of elevated waterborne arsenic concentrations in three counties classified by the study as having elevated waterborne arsenic concentrations (Pierce County, Washington; Sierra County, New Mexico; and Iberia County, Louisiana). Although we did not use these data in determining our estimates of mean arsenic exposure in counties, we examined the data for comparative purposes.

## CALCULATION OF COUNTY MEAN ARSENIC LEVELS

We chose to ascertain arsenic exposure levels by county rather than city or town, since mortality and cancer incidence data suitable for health effects studies are most commonly reported at the county level. Furthermore, there may be less misclassification of the location of residence of a cancer case if data are reported for a county rather than a city. For example, individuals residing close to the boundary of a city may be incorrectly reported to reside within that city. We determined the mean arsenic concentration of potential study counties in the following manner:

1. Using data we obtained from the state or the AOED when state data were not available, we estimated a mean arsenic concentration for each water system. An average arsenic concentration was estimated for each source by determining the

average value for all arsenic tests reported for that source. When we lacked individual well production data, we assumed that each source contributed an equal amount of drinking water to the utility. When well production data were available, arsenic concentrations were weighted by well production. In the absence of well production data, the initial estimate of the mean arsenic concentration for the utility was assumed to be the mean arsenic concentration for that utility.

2. For utilities with a single source or with multiple sources having similar arsenic concentrations (within 5 μg/L), the arsenic concentration for that utility was assumed to be the average arsenic concentration of the sources.
3. By aggregating our data into counties, we identified the counties likely to have a mean arsenic concentration exceeding 10 μg/L arsenic. A higher priority was given to improving the accuracy of the estimated mean arsenic concentration for water utilities in these counties. We sometimes obtained the mean arsenic exposure level for a utility from the utility Consumer Confidence Reports. In other cases, we obtained the information by directly contacting the utility. Most of our efforts were directed at obtaining drinking water arsenic concentrations for systems serving more than 10,000 people. A total of 54 utilities serving more than 10,000 people with elevated but uncertain drinking water arsenic concentrations were identified and we obtained arsenic concentrations from 43 (80%) of them. Seven of the 11 water systems that did not provide mean arsenic concentration data served fewer than 15,000 people. When we could not obtain better information we assumed the mean arsenic concentration for the water system was the mean concentration of the different sources.
4. Based on our best estimate of the water system mean water utility arsenic concentrations, we calculated the mean county arsenic concentration. Each water utility's mean arsenic concentration was weighted by the size of the population served. These weighted means were then summed for all water systems in the county to derive the estimated mean county arsenic concentration.
5. As a final check, after identifying the counties with a mean arsenic concentration of 10 μg/L or greater, we prepared a spreadsheet of all arsenic occurrence data from states, the AOED, the NRDC database, and water systems. This spreadsheet was checked to ensure that all data sources indicated elevated arsenic concentrations were observed for most water sources in the county and that the mean arsenic concentration was not greatly influenced by one or a small number of sources with very high arsenic concentrations. No outlying observations were identified in the counties with elevated arsenic concentrations. We also attempted to exclude all water systems that use surface-derived drinking water because surface water sources generally have much lower arsenic concentrations and because it is relatively easy to modify a surface water treatment plant to remove arsenic.

We excluded counties from further consideration if there were large differences in drinking water arsenic concentrations between water systems or if we could not account for most of the people in the county. The latter situation could occur if a large fraction of people were served by private wells or if arsenic data from one or more utilities were missing. We also excluded from further consideration any counties in which fewer than 75% of the public wells had elevated arsenic concentrations.

We identified all water systems with at least one well with an arsenic concentration above 10 μg/L using the AOED. The number of wells with elevated arsenic by state is shown in Table 3.2. In the 25 states contributing data to the AOED, there are 243 water systems with at least one well with arsenic concentrations above 50 μg/L. Of the wells exceeding 10 μg/L arsenic, 93% are in systems that serve fewer than 2,500 people and 98% are in systems that serve fewer than 10,000 people.

**Table 3.2**
**Number of wells with elevated arsenic**

| | Number of wells by arsenic concentration | | | | |
|---|---|---|---|---|---|
| State | 10-15 μg/L | 16-20 μg/L | 20-50 μg/L | 50 μg/L + | Total wells > 10 μg/L |
| Alaska | 98 | 44 | 104 | 23 | 269 |
| Alabama | 3 | 0 | 2 | 0 | 5 |
| Arizona | 114 | 40 | 80 | 27 | 261 |
| California | 211 | 100 | 165 | 54 | 530 |
| Colorado | 23 | 5 | 17 | 5 | 50 |
| Florida | 1 | 0 | 0 | 0 | 1 |
| Idaho | 35 | 17 | 29 | 2 | 83 |
| Illinois | 34 | 20 | 24 | 3 | 81 |
| Indiana | 5 | 1 | 1 | 0 | 7 |
| Kansas | 30 | 3 | 3 | 2 | 38 |
| Kentucky | 73 | 2 | 4 | 0 | 79 |
| Maine | 3 | 6 | 6 | 2 | 20 |
| Michigan | 425 | 255 | 418 | 53 | 1151 |
| Minnesota | 61 | 27 | 35 | 3 | 126 |
| Missouri | 8 | 6 | 3 | 0 | 17 |
| Montana | 8 | 4 | 5 | 0 | 17 |
| North Carolina | 53 | 9 | 26 | 6 | 94 |
| North Dakota | 16 | 8 | 14 | 1 | 39 |
| New Hampshire | 52 | 18 | 40 | 12 | 122 |
| New Jersey | 19 | 6 | 6 | 0 | 31 |
| New Mex. | 89 | 31 | 44 | 3 | 167 |
| Nevada | 48 | 27 | 57 | 26 | 158 |
| Ohio | 64 | 26 | 17 | 2 | 109 |
| Oklahoma | 47 | 21 | 27 | 9 | 104 |
| Oregon | 31 | 10 | 12 | 1 | 54 |
| Texas | 89 | 40 | 56 | 8 | 193 |
| Utah | 20 | 10 | 13 | 1 | 44 |
| All states | 1663 | 736 | 1208 | 243 | 3850 |

Table 3.3 shows the mean arsenic concentrations by county for our 'confirmed' counties, comparing data from various sources. Our estimated mean arsenic concentration for the county, derived according to the methods described above, are shown in the far right column of the table. Because of potential incompleteness of our data sources and the particular decision rules we applied in identifying counties for potential study, Table 3.2 may not include all U.S. counties with elevated drinking water arsenic levels.

**Table 3.3. Mean drinking water arsenic concentrations by data source for U.S. counties with a mean arsenic concentration of 10 µg/L or greater**

| State | County | State data (mean arsenic µg/L) * | AOED data (mean arsenic µg/L) † | Utility data (mean arsenic µg/L) ‡ | NRDC data (mean arsenic µg/L) §,†† | Engel and Smith, 1994 (mean arsenic µg/L), **,†† | Frost, estimated mean county arsenic concentration (µg/L) |
|---|---|---|---|---|---|---|---|
| Arizona | La Paz | NA | 11.7 | NA | 18.1 | | 11.7 |
| | Pinal | NA | 10.3 | NA | 13.9 | | 10.3 |
| California | Kings | 20.3 | 27.0 | 16.1 | 29.3 | 46.2 | 16.1 |
| | Mono | 19.3 | 37.1 | 13.0 | 15.1 | | 13.0 |
| Colorado | Alamosa | 36.9 | NA | NA | NA | 29.4 | 36.9 |
| | Lincoln | 23.4 | NA | NA | NA | 15.5 | 23.4 |
| | Rio Grande | 23.8 | NA | NA | NA | | 23.8 |
| Idaho | Payette | 14.4 | NA | NA | NA | | 14.4 |
| | Washington | 15.9 | NA | 17.0 | NA | 10.1 | 17.0 |
| Illinois | De Witt | NA | 17.1 | NA | 15.9 | | 17.1 |
| | Gallatin | 21.1 | 9.3 | 22.0 | 11.8 | 8.5 | 22.0 |
| North Dakota | Divide | NA | 13.6 | NA | 15.2 | | 13.6 |
| | Lamoure | NA | 14.9 | NA | 14.9 | | 14.9 |
| | Ramsey | NA | 17.9 | NA | 17.9 | | 17.9 |
| New Mexico | Bernalillo | 22.6 | 11.9 | 14.1 | 12.9 | | 14.1 |
| | Sandoval | 12.8 | 49.2 | 17.0 | 16.2 | | 17.0 |
| | Socorro | 28.1 | 35.1 | 32.2 | 22.3 | | 32.2 |
| Nevada | Churchill | 56.4 | 82.5 | 90.0 | 82.5 | | 90.0 |
| | Esmeralda | 14.8 | 25.6 | 36.0‡‡ | 30.5 | 91.5 | 25.6 |
| | Lander | 23.4 | 29.4 | NA | 15.6 | 24.6 | 17.3 |
| | Lincoln | 23.4 | 10.9 | 15.7 | 14.8 | 15.6 | 15.7 |
| | Lyon | 23.3 | 27.0 | 21.3 | 30.8 | 11.1 | 21.3 |
| | Nye | 8.8 | 13.4 | 13.6 | 18.3 | 35.9 | 13.6 |

(continued)

**Table 3.3 (Continued)**

| State | County | State data (mean arsenic µg/L) * | AOED data (mean arsenic µg/L) † | Utility data (mean arsenic µg/L) ‡ | NRDC data (mean arsenic µg/L ) §,†† | Engel and Smith, 1994 (mean arsenic µg/L), **,†† | Frost, estimated mean county arsenic concentration ( µg/L) |
|---|---|---|---|---|---|---|---|
| Oklahoma | Canadian | 19.5 | 12.4 | 26.9 | NA | 10.0 | 26.9 |
| | Custer | 13.0 | 12.7 | 31.0‡‡ | NA | 19.3 | 13.0 |
| Texas | Andrews | 35.6 | 33.7 | NA | 34.6 | 8.0 | 33.6 |
| | Borden | 22.0 | 22.6 | NA | 22.6 | | 22.0 |
| | Gaines | NA | 12.1 | NA | 20.3 | | 12.1 |
| | Hudspeth | NA | 11.6 | NA | 14.5 | | 11.6 |
| | Jim Hogg | 54.0 | 88.5 | 77.9 | 33.3 | | 77.9 |
| | Karnes | 15.6 | 45.5 | NA | 33.9 | | 15.6 |
| | Yoakum | NA | 11.7 | NA | 11.4 | | 11.7 |
| Utah | Summit | NA | 12.6 | NA | 17.5 | | 12.6 |

*Collected directly from state agencies by author

†EPA (Environmental Protection Agency). 2000. *Arsenic Occurrence in Public Drinking Water Supplies*. EPA-815-R-00023. Washington, D.C.: U.S. Environmental Protection Agency.

‡Collected directly from water utilities by author

§Mushak, M. 2000. *Arsenic and Old Laws. A Scientific and Public Health Analysis of Arsenic Occurrence in Drinking Water, its Health effects, and EPA's Outdated Arsenic Tap Water Standard*. New York, NY: National Resources Defense Council.

**Engel, R.R., and A.H. Smith. 1994. Arsenic in Drinking Water and Mortality From Vascular Disease: An Ecologic Analysis in 30 Counties in the United States. *Archives of Environmental Health*, 49(5):418-27.

††Data presented for comparison purposes only; not used by author for calculating county arsenic exposure estimates.

‡‡Not all utilities in the county were contacted

NA--data not available

## Population Covered by Captured Arsenic Exposure Data

To ensure that the water systems used in our estimates accounted for most of that county's population, we compared the population served as estimated by the utilities in the county to the county's 1990 population as estimated by the U.S. Census. These population size comparisons are presented in Table 3.4. The size of the population served by a water system was an estimate provided by the utility operator, and may not have been accurate in all instances. There appeared to be over-counting by some water systems, in that data indicate the number of people served was greater than the number of people residing in the county in 1990. In general, however, this comparison of population sizes did not reveal any significant gaps in our data capture, with the exception of drinking water systems in Nevada for which the size of the population served was not available.

**Table 3.4**
**Counties with arsenic concentrations 10 μg/L or greater**

| State | County | 1990 population (U.S. Census) | Population covered by utility arsenic data | Frost, estimated mean arsenic concentration ( μg/L) |
|---|---|---|---|---|
| Arizona | La Paz* | 13,844 | 10,326 | 11.7 |
| | Pinal | 116,379 | 154,247 | 10.3 |
| California | Kings | 101,469 | 96,111 | 16.1 |
| | Mono | 9,956 | 7,561 | 13.0 |
| Colorado | Alamosa | 13,617 | 11,865 | 36.9 |
| | Lincoln | 4,529 | 2,880 | 23.4 |
| | Rio Grande | 10,770 | 6,396 | 23.8 |
| Idaho | Payette | 16,434 | 10,190 | 14.4 |
| | Washington | 8,550 | 6,210 | 17.0 |
| Illinois | De Witt | 17,039 | 12,327 | 17.1 |
| | Gallatin | 6,909 | 5,780 | 22.0 |
| North Dakota | Divide | 2,899 | 1,596 | 13.6 |
| | Lamoure | 5,383 | 1,737 | 14.9 |
| | Ramsey | 12,681 | 9,795 | 17.9 |
| New Mexico | Bernalillo | 480,577 | 452,960 | 14.1 |
| | Sandoval | 63,319 | 60,000 | 17.0 |
| | Socorro | 14,764 | 13,964 | 32.2 |
| Nevada | Churchill | 17,938 | NA | 90.0 |
| | Esmeralda | 1,344 | NA | 25.6 |
| | Lander | 6,266 | NA | 17.3 |
| | Lincoln | 3,775 | NA | 15.7 |
| | Lyon | 20,001 | NA | 21.3 |
| | Nye | 17,781 | NA | 13.6 |
| Oklahoma | Canadian | 74,409 | 44,633 | 26.9 |
| | Custer | 26,897 | 33,890 | 13.0 |

(continued)

**Table 3.4 (Continued)**

| State | County | 1990 population (U.S. Census) | Population covered by utility arsenic data | Frost, estimated mean arsenic concentration ( µg/L) |
|---|---|---|---|---|
| Texas | Andrews | 14,338 | 11,446 | 33.6 |
| | Borden | 799 | 190 | 22.0 |
| | Gaines | 14,123 | 9,150 | 12.1 |
| | Hudspeth | 2,915 | 3,080 | 11.6 |
| | Jim Hogg | 5,109 | 24,536 | 77.9 |
| | Karnes | 12,455 | 25,218 | 15.6 |
| | Yoakum | 8,786 | 6,752 | 11.7 |
| Utah | Summit | 15,518 | 17,440 | 12.6 |

*La Paz County was not included as an arsenic-exposed county in the analysis; for reasons described below, we incorporated it into Yuma County, a comparison county.

NA—Data not available

## PERSON-YEARS OF EXPOSURE

Table 3.5 summarizes the persons-years of arsenic exposure for each decade (1950-59, 1960-69, 1970-79, 1980-89, 1990-99) for all 'confirmed' counties with arsenic concentrations greater than 10 µg/L, 20 µg/L, and 50 µg/L. Similar data by age are shown in Table 3.6. In the United States for the 50-year period 1950-1999, there are over 51 million person-years of exposure to waterborne arsenic exceeding 10 µg/L, 8.8 million person-years of exposure exposed to waterborne arsenic exceeding 20 µg/L arsenic, and 0.9 million person-years of exposure to waterborne arsenic exceeding 50 µg/L arsenic.

**Table 3.5**
**Person-years of exposure by decade for counties, with >10 µg/L arsenic exposure**

| | Person-Years | | |
|---|---|---|---|
| | Waterborne arsenic >10 µg/L | Waterborne arsenic >20 µg/L | Waterborne arsenic >50 µg/L |
| 1990-99 | 13,671,915 | 2,168,230 | 254,455 |
| 1980-89 | 12,279,201 | 1,990,602 | 208,679 |
| 1970-79 | 10,282,575 | 1,662,076 | 169,301 |
| 1960-69 | 8,153,019 | 1,293,488 | 142,359 |
| 1950-59 | 6,716,000 | 1,072,255 | 124,158 |
| Total | 51,102,709 | 8,168,651 | 921,999 |
| Mean arsenic level ( µg/L) | 17.5 | 35.1 | 87.9 |

**Table 3.6**
**Person-years of exposure by decade, age and arsenic exposure group**

| Time period | Arsenic exposure group | Number of counties | Person-years age 0-19 | Person- years age 20-49 | Person- years age 50-74 | Person- years age 75+ | Person-years all ages |
|---|---|---|---|---|---|---|---|
| 1950-99 | ≤10 µg/L | 652 | 361,154,774 | 426,185,869 | 215,142,317 | 47,230,458 | 1,049,713,418 |
| 1950-99 | 11-19 µg/L | 20 | 11,675,985 | 14,104,617 | 6,315,987 | 1,230,143 | 33,326,732 |
| 1950-99 | ≥20 µg/L | 12 | 2,385,067 | 2,725,741 | 1,347,937 | 276,601 | 6,735,346 |
| 1950-59 | ≤10 µg/L | 652 | 59,903,565 | 62,193,329 | 30,135,904 | 4,599,206 | 156,832,004 |
| 1950-59 | 11-19 µg/L | 20 | 1,768,090 | 1,739,746 | 668,376 | 85,209 | 4,261,421 |
| 1950-59 | ≥20 µg/L | 12 | 387,549 | 369,845 | 178,019 | 26,024 | 961,437 |
| 1960-69 | ≤10 µg/L | 652 | 69,029,657 | 65,136,847 | 35,857,936 | 6,449,866 | 176,474,306 |
| 1960-69 | 11-19 µg/L | 20 | 2,231,319 | 2,021,394 | 893,586 | 132,833 | 5,279,132 |
| 1960-69 | ≥20 µg/L | 12 | 413,255 | 375,663 | 201,260 | 36,080 | 1,026,258 |
| 1970-79 | ≤10 µg/L | 652 | 72,954,510 | 78,975,481 | 43,091,557 | 8,571,827 | 203,593,375 |
| 1970-79 | 11-19 µg/L | 20 | 2,414,412 | 2,594,567 | 1,231,764 | 198,866 | 6,439,609 |
| 1970-79 | ≥20 µg/L | 12 | 453,987 | 478,316 | 246,771 | 49,088 | 1,228,162 |
| 1980-89 | ≤10 µg/L | 652 | 76,074,632 | 102,223,638 | 49,856,677 | 11,844,268 | 239,999,215 |
| 1980-89 | 11-19 µg/L | 20 | 2,466,524 | 3,514,334 | 1,563,805 | 317,769 | 7,862,432 |
| 1980-89 | ≥20 µg/L | 12 | 540,144 | 696,230 | 321,476 | 69,974 | 1,627,824 |
| 1990-99 | ≤10 µg/L | 652 | 83,192,410 | 117,656,574 | 56,200,243 | 15,765,291 | 272,814,518 |
| 1990-99 | 11-19 µg/L | 20 | 2,795,640 | 4,234,576 | 1,958,456 | 495,466 | 9,484,138 |
| 1990-99 | ≥20 µg/L | 12 | 590,132 | 805,687 | 400,411 | 95,435 | 1,891,665 |

## STATISTICAL POWER

Background U.S. mortality rates for lung and bladder cancer combined range from 27 per 100,000 in females for 1950-59 and increase to 49 per 100,000 in females by 1970-79. Considering only arsenic-related cancers in females and assuming a baseline mortality rate 40 per 100,000 for the period 1950-99, a power of 80%, and a significance level of 5%, we estimate that 450,000 person-years of exposure is more than sufficient to detect an increased relative risk of 1.25 for these two cancers among people exposed to 50 μg/L or more drinking water arsenic. These calculations assume a straightforward comparison and not adjustment for neighboring county characteristics (i.e., multi-level modeling). Therefore, they provide a conservative estimate of the true power of the study to detect an effect. If the latency between completion of exposure and cancer diagnosis is greater than 40 years these power estimates will over-estimate the true power of the study.

Sample sizes for specific years (1955-1975) for males and females with a range of relative risks of arsenic–associated cancer death are given in Table 3.7. This table should be considered to be a general guide only since the population size needed to detect an anticipated effect also depends on the level of in-migration and the net out-migration for age groups of interest, as discussed in Chapter 2. The level of migration increases for longer latency periods, reducing the power of the study to detect an arsenic-related health effect. Since people who leave an area may also return later, the net out-migration rate is the difference between the out-migration rates and the return migration rate.

**Table 3.7**
**Sample size needed to detect a given relative risk (RR)* (alpha = 0.05, power 0.90, ratio unexposed to exposed = 4)**

| Year | Unexposed rate, males | Sample size needed | Unexposed rate, females | Sample size needed |
|---|---|---|---|---|
| RR 1.25 | | | | |
| 1955 | 35.1 | 678,521 | 8.0 | 2,993,891 |
| 1965 | 50.1 | 479,633 | 10.8 | 2,217,626 |
| 1975 | 69.2 | 345,872 | 20.7 | 1,146,466 |
| RR 1.50 | | | | |
| 1955 | 35.1 | 190,149 | 8.0 | 838,821 |
| 1965 | 50.1 | 133,395 | 10.8 | 621,326 |
| 1975 | 69.2 | 96,897 | 20.7 | 321,352 |
| RR 1.75 | | | | |
| 1955 | 35.1 | 94,144 | 8.0 | 412,548 |
| 1965 | 50.1 | 65,761 | 10.8 | 305,579 |
| 1975 | 69.2 | 47,651 | 20.7 | 159,850 |

* CDC 2000. EpiInfo. Sample Size, Uitenbroek 1997. http://home.clara.net/sisa/sampshlp.htm

## COUNTIES IN FINAL STUDY COHORT

### Arsenic-Exposed Counties

The arsenic-exposed counties that we identified using the methods described above were located in the following states: Arizona, California, Colorado, Idaho, Illinois, North Dakota, New

Mexico, Nevada, Oklahoma, Texas, and Utah. We identified 33 out of 744 counties in these 11 states in which the average arsenic concentration of at least 75% of public wells exceeded 10 μg/L (Table 3.4). Of these 33 'confirmed' counties, 12 had an average arsenic concentration that exceeded 20 μg/L and 2 had an average arsenic concentration that exceeded 50 μg/L arsenic.

**Selection of Comparison Counties**

We next selected the comparison counties required for the analysis. Since geographic locations frequently share common socio-economic characteristics as well as topographical features, all remaining counties in the 11 states (N = 711) potentially served as the comparison counties in the statistical model. A comparison county was excluded if it met either of the following two criteria:

1. The county was part of a primary metropolitan statistical area (PMSA; population area of 1,000,000 with "substantial commuting interchange").
2. The county had a 1990 U.S. Census population of over 1,000,000.

This step resulted in the exclusion of 59 comparison counties. These criteria were applied for three reasons. First, because none of the arsenic-exposed counties had populations of 1,000,000 or greater, it would not be appropriate to have very large counties as comparison areas. Secondly, many large counties have industries and exposures that smaller counties do not have. Thirdly, although we transformed the data to reduce the effect of population size on the analysis, the addition of counties with very large populations might have skewed the analyses simply because of the population size. Since the comparison would have provided no benefit to the analysis, these counties were excluded.

Due to the longitudinal nature of our study, as a final step we aggregated into one county those counties that had separated during the study time frame. La Paz County, Arizona (an arsenic-exposed county) was formed from the northern part of Yuma County (a comparison county) in 1983. In our study, La Paz was considered part of Yuma County. In 1981, Valencia County in New Mexico was divided into Valencia County and Cibola County. In our study, Cibola County (a potential comparison county) was considered part of Valencia County (also a comparison county).

The final cohort thus comprised 684 counties from 11 states. Of these counties, 32 were arsenic-exposed (mean drinking water arsenic >10 μg/L) and 650 were non-exposed comparison counties.

# CHAPTER 4
# METHODS PART 3: STATISTICAL ANALYSIS

## ANALYTICAL MODELS

In separate models, we examined whether excess bladder cancer mortality, bladder cancer incidence, lung cancer mortality, and lung cancer incidence occurs in populations exposed to high levels of arsenic in drinking water. We used SAS statistical software, version 8.2 (SAS Institute Inc., Cary, North Carolina) to aggregate data for analysis. To perform multivariate modeling, we used MLwiN statistical software, version 2.0 (Centre for Multi-level Modeling, Institute of Education, London, England).

We developed multi-level models to determine if counties with elevated drinking water arsenic levels have higher rates of either lung or bladder incidence or mortality than neighboring areas. We evaluated mortality risk using three different approaches. In the first approach, we combined all cancer deaths across the decades for which data were available, 1950-1999. In the second approach, essentially a sub-analysis of the first, we limited the analysis to the population age 50 years and older. In the third approach, we more precisely adjusted for socioeconomic characteristics of the counties. For this analysis, we combined cancer deaths by decade, using only decades for which comparable socioeconomic variables were available (1960-1999); this was therefore a repeated measure analysis of decades within county. This third analysis more precisely adjusted for the covariate effect of socioeconomic variables on mortality rates, as we used the socioeconomic characteristics of the midpoint of each decade. In the previous combined decades models, rankings of counties with respect to socioeconomic covariates were assumed to remain constant across the 50-year time period. The socioeconomic variables for one decade, the 1980s, were used as a measure of a county's status with respect to other counties for all the decades. Cancer incidence analyses were accomplished by combining individual year data from 1973-1999 into three decades; as in the third mortality analysis approach, this was a repeated measure analysis of decades within county.

Where possible, analyses were performed using a continuous, normally distributed response variable. For the cancer mortality analyses, the response variable was the age-standardized mortality ratio (SMR). For the cancer incidence analyses, the response variable was the age-standardized incidence ratio (SIR). In modeling rarer occurrences, such as bladder cancer or lung cancer within individual decades, where low values of expected occurrences are a source of more variation in the calculated ratios, actual counts of occurrences are modeled using a Poisson response model. The response variable was the observed number of occurrences. Poisson response models incorporate a logarithmic function and an adjustment for expected occurrences. Covariate information is included in all models for area-specific attributes and population demographics.

Hierarchical models are multi-level models of counties within neighboring areas. Additionally, some models incorporate a repeated measures component (counts of occurrences for distinct time periods). In these cases, the hierarchical models are multi-level models of time periods within counties within neighboring areas.

Hierarchical models are random intercept, fixed covariate effect models. Model parameters have been estimated using restricted iterative generalized least squares estimation (RIGLS) and Markov Chain Monte Carlo Methods (MCMC). RIGLS models were run before MCMC models to provide initial estimates for model parameters.

## Model Levels

The unit of analysis in our study is the individual county. In our models we investigated the degree to which counties exhibited variation; however neighboring areas also can be an influential factor in outcome variation. Therefore, the level above county in our model was the neighboring area around the county.

Identification of proximate, or neighboring, counties was accomplished using two different methods. In the first method a county was designated as neighboring if the counties shared a border. Because of the arbitrary nature of county boundaries and the differences in land mass in counties between states (i.e., Illinois and Texas have small size counties compared to Arizona and Nevada), we also used a second method in which a county was designated as neighboring if it was within twice the mean minimum distance between any two county centroids in the study. The centroid of a county is the geographical midpoint of the county based on Cartesian coordinate data. SAS statistical software version 8.02 was used to identify the neighboring counties. Using the first definition of neighboring county, the mean number of neighboring counties was 6.4 and the maximum was 14. For the second definition, the mean number of neighboring counties was 8.8 and the maximum was 23. Neighborhood effects were investigated in most models using the first definition of neighboring county. As a validation test of the neighborhood effect, the second definition was also applied to models investigating cancer mortality when all decades were combined and the population was the entire county population.

# HEALTH OUTCOMES DATA USED IN MODELS

## Cancer Mortality Data

Mortality data (bladder and lung cancer) for study counties for the periods 1950-59, 1960-69, 1970-79, 1980-89, 1990-99, and 1950-1999 were obtained from NCI. Data used are for white populations. Because of the small size of black and other non-white populations in many Western state counties, mortality for these races/ethnicities was not considered. Data were categorized by sex and time period for the four age groups described above. The data also include population counts (denominators) for each category (age, sex, time period). These denominators were used to calculate the age-adjusted SMRs for each county for each sex and time period. (See Tables 4.1 and 4.2).

**Table 4.1**
**Bladder cancer mortality 1950-1999 within study cohort by sex and age**

| | Deaths per 100,000 (number of deaths) | | | |
|---|---|---|---|---|
| | Females | | Males | |
| Years | All ages | Age 50+ | All ages | Age 50+ |
| 1950-1999 | 2.23 (12,229) | 8.18 (11,935) | 5.26 (28,482) | 22.0 (27,758) |
| 1950-1959 | 2.08 (1,675) | 8.86 (1,592) | 4.56 (3,727) | 20.0 (3,559) |
| 1960-1969 | 2.11 (1,949) | 8.23 (1,882) | 5.27 (4,789) | 22.0 (4,663) |
| 1970-1979 | 2.23 (2,388) | 8.00 (2,326) | 5.54 (5,784) | 23.1 (5,657) |
| 1980-1989 | 2.16 (2,735) | 7.61 (2,678) | 5.04 (6,202) | 21.1 (6,092) |
| 1990-1999 | 2.47 (3,552) | 8.46 (3,457) | 5.66 (7,980) | 22.8 (7,787) |

**Table 4.2**
**Lung cancer mortality 1950-1999 within study cohort by sex and age**

| Years | Deaths per 100,000 (number of deaths) | | | |
|---|---|---|---|---|
| | Females | | Males | |
| | All ages | Age 50+ | All ages | Age 50+ |
| 1950-1999 | 22.1 (121,854) | 78.3 (114,153) | 54.1 (292,577) | 220.0(276,707) |
| 1950-1959 | 4.5 (3,646) | 17.4 (3,135) | 22.1 (18,119) | 90.1 (15,971) |
| 1960-1969 | 7.2 (6,610) | 24.6 (5,626) | 39.1 (35,564) | 156.8 (32,476) |
| 1970-1979 | 15.5 (16,596) | 51.2 (14,833) | 56.7 (59,140) | 227.1 (55,410) |
| 1980-1989 | 28.2 (35,718) | 95.8 (33,670) | 66.9 (82,339) | 273.8 (78,929) |
| 1990-1999 | 41.3 (59,284) | 139.4 (56,889) | 69.2 (97,415) | 275.5 (93,921) |

### Cancer Incidence Data

NCI Surveillance, Epidemiology, and End Results (SEER) cancer incidence and associated population data for the period 1973-1999 from the New Mexico and Utah registries were used in calculating age-adjusted SIRs for each county for each sex and time period. As with the mortality data, we calculated SIRs only for white populations. Incidence among black and nonwhite populations was not considered (see Tables 4.3 and 4.4).

**Table 4.3**
**Bladder cancer incidence 1973-1999 by sex**

| Years | Incidence per 100,000 | |
|---|---|---|
| | Females | Males |
| 1973-1979 | 5.12 (419) | 15.77(11,265) |
| 1980-1989 | 5.51 (782) | 18.56 (2,583) |
| 1990-1999 | 6.73 (1,119) | 21.49 (3,504) |

**Table 4.4**
**Lung cancer incidence 1973-1999 by sex**

| Years | Incidence per 100,000 | |
|---|---|---|
| | Females | Males |
| 1973-1979 | 13.34 (1,092) | 38.19 (3,063) |
| 1980-1989 | 19.84 (2,818) | 41.06 (5,715) |
| 1990-1999 | 26.50 (4,403) | 41.91 (6,831) |

### Standardized Mortality Ratios (SMR)

For each sex in each time period, the expected number of cancer deaths for each county was determined from the age distribution of the county population. The observed deaths, the population (person-years for the time period), and the expected number of deaths were reported for all ages and for the following age groups:

1. 0-19 years of age
2. 20-49 years of age
3. 50-74 years of age
4. ≥ 75 years of age

An expected number of cancer deaths for each of these age groups (a = 1 to 4) for each of the counties (i = 1 to 684) is then:

$$\text{Expected}_{i\,a} = \text{Population}_{i\,a} \times (\text{total cancers*}_{a} / \text{total population*}_{a})$$

* across all counties with reference level of arsenic ≤10 μg/L, for sex and decade

The expected number of cancer deaths for a county (i = 1 to 684) for the given decade and sex group is then:

$$\text{Expected}_{i} = \text{SUM}(\text{Expected}_{i\,a}) \text{ where a= age groups}$$

With the standardized mortality ratio (SMR) then:

$$\text{SMR}_{i} = \text{Observed}_{i} / \text{Expected}_{i}, \text{ where Observed}_{i} = \text{SUM}(\text{Observed}_{i\,a})$$

The calculated SMRs for bladder and lung cancer for all individual decades and all decades combined are shown in Tables 4.5–4.7. There is sizable variation in calculated SMRs for bladder cancer due to the low bladder cancer mortality rate, particularly among females. This is also true for lung cancer in females in the earlier decades. This large variation is an indication that the problems related to small area populations and rare events with respect to a calculated ratio that were discussed in Chapter 2 could be problems here. Conversely, there is not the same extreme variation in lung cancer SMRs across the five-decade period 1950-1999. For these reasons, we used a Poisson distribution model rather than the continuous SMR for modeling bladder cancer mortality for individual decades (Table 4.5), bladder cancer for all decades (Table 4.7), and lung cancer for individual decades (Table 4.6). Lung cancer mortality for all decades combined was modeled using SMR as the response variable (Table 4.7).

**Table 4.5**
**Bladder cancer standardized mortality ratio (SMR) by decade and drinking water arsenic level**

| Decade | ≤ 10 μg/L (n=652) | 11 – 19 μg/L (n=20) | ≥ 20 μg/L (n=12) |
|---|---|---|---|
| | | Females SMR<br>Mean (SE) [min,max] | |
| 1960s | 0.92 (0.07) [0,35.81] | 0.63 (0.17) [0,2.57] | 0.19 (0.13) [0,1.31] |
| 1970s | 0.90 (0.04) [0,10.55] | 0.66 (0.15) [0,1.80] | 1.17 (0.33) [0,3.56] |
| 1980s | 0.90 (0.03) [0,6.70] | 0.93 (0.22) [0,3.19] | 1.87 (0.78) [0,10.21] |
| 1990s | 0.92 (0.03) [0,5.70] | 0.84 (0.12) [0,1.87] | 1.34 (0.98) [0,12.10] |
| | | Total Observed Cancer Deaths | |
| 1960s | 1,909 | 38 | 2 |
| 1970s | 2,313 | 59 | 16 |
| 1980s | 2,634 | 83 | 18 |
| 1990s | 3,443 | 95 | 14 |

(continued)

**Table 4.5 (Continued)**

| Decade | ≤ 10 μg/L (n=652) | 11 – 19 μg/L (n=20) | ≥ 20 μg/L (n=12) |
|---|---|---|---|
| | | Males SMR Mean (SE) [min,max] | |
| 1960s | 0.88 (0.03) [0,5.00] | 0.92 (0.15) [0,2.50] | 1.01 (0.26) [0,3.43] |
| 1970s | 0.92 (0.03) [0,7.58] | 0.88 (0.13) [0,2.07] | 1.10 (0.28) [0,3.23] |
| 1980s | 0.92 (0.02) [0,5.30] | 0.78 (0.10) [0,1.44] | 0.95 (0.21) [0,2.66] |
| 1990s | 0.96 (0.03) [0,5.18] | 0.83 (0.12) [0,1.82] | 0.91 (0.24) [0,2.29] |
| | | Total Observed Cancer Deaths | |
| 1960s | 4,653 | 111 | 25 |
| 1970s | 5,587 | 158 | 39 |
| 1980s | 5,999 | 169 | 34 |
| 1990s | 4,698 | 226 | 56 |

**Table 4.6**
**Lung cancer standardized mortality ratio (SMR) by decade and drinking water arsenic level**

| Decade | ≤ 10 μg/L (n=652) | 11 – 19 μg/L (n=20) | ≥ 20 μg/L (n=12) |
|---|---|---|---|
| | | Females SMR Mean (SE) [min,max] | |
| 1960s | 0.95 (0.03) [0,9.60] | 1.00 (0.18) [0,3.50] | 0.80 (0.16) [0,1.55] |
| 1970s | 0.86 (0.02) [0,4.61] | 0.94 (0.09) [0.40,1.87] | 0.94 (0.11) [0,1.72] |
| 1980s | 0.86 (0.02) [0,3.39] | 0.86 (0.07) [0.29,1.40] | 1.04 (0.17) [0.35,2.52] |
| 1990s | 0.90 (0.02) [0,5.88] | 0.88 (0.06) [0.23,1.34] | 1.04 (0.12) [0.47,1.81] |
| | | Total Observed Cancer Deaths | |
| 1960s | 6,409 | 170 | 31 |
| 1970s | 16,097 | 416 | 83 |
| 1980s | 34,590 | 917 | 211 |
| 1990s | 57,201 | 1,676 | 407 |
| | | Males SMR Mean (SE) [min,max] | |
| 1960s | 0.90 (0.01) [0,2.16] | 0.76 (0.08) [0.15,1.54] | 0.88 (0.09) [0.32,1.44] |
| 1970s | 0.92 (0.01) [0,1.97] | 0.80 (0.07) [0.22,1.35] | 0.84 (0.09) [0.38,1.38] |
| 1980s | 0.93 (0.01) [0,1.92] | 0.84 (0.04) [0.37,1.10] | 0.82 (0.10) [0.44,1.52] |
| 1990s | 1.00 (0.01) [0,2.43] | 0.82 (0.06) [0.29,1.39] | 0.90 (0.12) [0.44,1.89] |
| | | Total Observed Cancer Deaths | |
| 1960s | 34,677 | 717 | 170 |
| 1970s | 57,539 | 1,282 | 319 |
| 1980s | 79,899 | 1,959 | 481 |
| 1990s | 94,240 | 2,561 | 624 |

**Table 4.7**
**Bladder and lung cancer standardized mortality ratios (SMR) 1950-1999 by arsenic level**

| | Drinking water arsenic level | | |
|---|---|---|---|
| | < 10 μg/L (n=652) | 11 – 19 μg/L (n=20) | > 20 μg/L (n=12) |
| | | Bladder cancer<br>Mean SMR (SE) [min,max] | |
| Females | 0.91 (0.02) [0,4.37] | 0.80 (0.09) [0,1.34] | 1.15 (0.23) [0.38,2.74] |
| Males | 0.91 (0.01) [0,1.90] | 0.83 (0.07) [0,1.38] | 0.86 (0.11) [0,1.26] |
| | | Total Observed Cancer Deaths | |
| Females | 11,933 | 306 | 60 |
| Males | 27,591 | 728 | 163 |
| | | Lung cancer<br>Mean SMR (SE) [min,max] | |
| Females | 0.89 (0.01) [0,2.10] | 0.89 (0.05) [0.27,1.34] | 1.01 (0.08) [0.56,1.46] |
| Males | 0.94 (0.01) [0,1.64] | 0.81 (0.04) [0.42,1.18] | 0.85 (0.08) [0.56,1.46] |
| | | Total Observed Cancer Deaths | |
| Females | 117,842 | 3,265 | 747 |
| Males | 284,020 | 6,870 | 1,687 |

**Standardized Incidence Ratios (SIR)**

We used similar methods for computing the SIR, but with differences in the number of age groupings and the number of counties. Because limited cancer incidence data were available for the 11 states in the broader study, only counties in New Mexico and Utah (states with SEER cancer registries), were included in this analysis.

The population (person-years for the time period) in each county was categorized into the nineteen SEER*Stat age groups. These comprise seventeen 5-year interval groupings for ages 1-84, plus the age groupings < 1 year and ≥ 85 years.

An expected number of cancer cases for each of these age groups (a = 1 to 19) for each of the counties (i = 1 to 61) is then:

$$\text{Expected}_{ia} = \text{Population}_{ia} \times (\text{ total cancers*}_a / \text{total population*}_a)$$

* across all counties with reference level of arsenic ≤10 μg/L, for sex and decade

The expected number of cancer cases for a county (i = 1 to 61) for the given decade and sex group is then:

$$\text{Expected}_i = \text{SUM}(\text{Expected}_{ia}) \text{ where a= age groups}$$

With the standardized incidence ratio (SIR) then:

$$\text{SIR}_i = \text{Observed}_i / \text{Expected}_i\text{, where Observed}_i = \text{SUM}(\text{Observed}_{ia})$$

Calculated SIRs for bladder and lung cancer incidence are shown in Table 4.8 and 4.9. As was true for calculated bladder cancer SMRs, there is a sizable amount of variation in county bladder cancer SIRs as opposed to county lung cancer SIRs. Thus bladder cancer incidence models were Poisson distribution models of incidence while lung cancer incidence models were continuous response models of county SIRs.

**Table 4.8**
**Bladder cancer SIR by decade and arsenic level**

| Decade | Drinking water arsenic level ≤10 μg/L (n=57) | | >10 μg/L (n=4) | |
|---|---|---|---|---|
| | **Females** | | | |
| | Mean SIR (SE) [min,max] | Total Observed Cancers | Mean SIR (SE) [min,max] | Total Observed Cancers |
| 1970s | 0.93 (0.14) [0,6.40] | 328 | 1.21 (0.18) [0.80,1.66] | 91 |
| 1980s | 0.96 (0.06) [0,2.24] | 598 | 1.66 (0.17) [1.29,2.08] | 184 |
| 1990s | 0.97 (0.09) [0,4.66] | 873 | 1.26 (0.14) [1.03,1.66] | 246 |
| | **Males** | | | |
| | Mean SIR (SE) [min,max] | Total Observed Cancers | Mean SIR (SE) [min,max] | Total Observed Cancers |
| 1970s | 0.81 (0.05) [0,1.86] | 1,027 | 1.25 (0.06) [1.13,1.42] | 238 |
| 1980s | 0.90 (0.05) [0,1.75] | 2,068 | 1.11 (0.09) [0.83,1.26] | 515 |
| 1990s | 0.95 (0.05) [0,1.83] | 2,839 | 0.92 (0.17) [0.48,1.30] | 665 |

**Table 4.9**
**Lung cancer standardized incidence ratio (SIR) by decade and drinking water arsenic level**

| Decade | Drinking water arsenic level ≤10 μg/L (n=57) | | >10 μg/L (n=4) | |
|---|---|---|---|---|
| | **Females** | | | |
| | Mean SIR (SE) [min,max] | Total Observed Cancers | Mean SIR (SE) [min,max] | Total Observed Cancers |
| 1970s | 1.01 (0.09) [0,2.82] | 824 | 0.88 (0.26) [0.31,1.57] | 268 |
| 1980s | 1.01 (0.06) [0,2.00] | 2111 | 1.19 (0.24) [0.49,1.48] | 707 |
| 1990s | 1.06 (0.06) [0,1.99] | 3295 | 1.16 (0.23) [0.50,1.52] | 1108 |
| | **Males** | | | |
| | Mean SIR (SE) [min,max] | Total Observed Cancers | Mean SIR (SE) [min,max] | Total Observed Cancers |
| 1970s | 0.93 (0.06) [0,2.23] | 2439 | 1.04 (0.15) [1.13,1.31] | 624 |
| 1980s | 1.02 (0.06) [0,2.56] | 4545 | 1.01 (0.12) [0.83,1.20] | 1170 |
| 1990s | 1.06 (0.05) [0,2.01] | 5382 | 1.14 (0.11) [0.48,1.38] | 1449 |

While bladder cancer SIRs for females were elevated in all decades, it would be misleading to draw conclusions based solely on these SIRs. There are only four counties in the higher arsenic group, one of which is a sizable metropolitan area in New Mexico; the only other sizable metropolitan area is in the lower arsenic group and is in Utah. Two of the other four counties in the higher arsenic group had expected bladder cancer incidences of less than 5 for each decade. Observed occurrences in these two counties did not exceed expected occurrences by more than two, but as discussed in Chapter 2, use of these low numbers in calculating ratios is not informative. In the fourth county, expected occurrences ranged from 3 to 20, with calculated SIRs ranging from 1.05 to 2.08.

## EXPLANATORY COVARIATE VARIABLES

### Metropolitan Area

Counties were designated as metropolitan if they were identified as a metropolitan statistic area (MSA) by the Federal Office of Management and Budget (OMB). An area qualifies as an MSA if it includes a city with 50,000 or more inhabitants or includes an Urbanized Area (UA) and has a total population of at least 100,000 (or 75,000 in New England). (http://www.census.gov/main/www/cen2000.html)

### Socioeconomic Variables

County-specific census information was used as a measure of each county's average social and economic status at designated time periods. The census items, their definitions, and source data are shown in Table 4.10. In order to negate any time effects inherent in crude numbers, model variables were normalized to the mean value for the designated time period. Census items are those that in preliminary analyses were associated with bladder and lung mortality or incidence.

**Table 4.10**
**Census variables used in statistical models**

| Description | 1990-1999 | 1980-1989 | 1970-1979 | 1960-1969 |
|---|---|---|---|---|
| Percent employed in manufacturing | Manufacturing employees 1995/all industry employees 1995* | Manufacturing employees 1985/all industry employees 1985* | Manufacturing employees 1979/all industry employees 1979* | % employed in manufacturing 1967 ‡ |
| Percentage of persons age ≥25 years with at least a college degree | Population age 25 years and over, 2000; percent with bachelor's degree or higher † | Sum (persons age 25 years and older completing high school only 1990, persons 25 years and over completing 1-3 years college 1990, persons 25 years and older completing 4 years or more of college 1990)/persons 25 years and older 1990* | Sum (persons age 25 years and older completing high school only 1980, persons 25 years and over completing 1-3 years college 1980, persons 25 years and older completing 4 years or more of college 1980)/persons 25 years and older 1980* | Percentage age 25 years and older; percent with bachelor's degree or higher, 1970 ‡ |
| Persons per household | Persons in households, 2000: persons per household† | Persons per household 1990* | Persons per household 1980* | Persons per household 1970* |
| Farm land as percentage of total county land | Farm land in acres 1992/(1990 land in square miles x 640 acres/mile)* | Farm land in acres 1982 / (1980 land in square miles x 640 acres/mile)* | Farm land 1974/(total land 1970 x 640 acres/mile) § | Farm land 1964/(total land 1960 x 640 acres/mile) § |
| Personal per capita income (1,000 $) | Per capita income in 1999 † | Per capita income in 1989** | Per capita income in 1979** | Per capita income in 1969** |

*U.S. Dept. of Commerce, Bureau of the Census. *USA Counties 1998*. Washington, D.C.: U.S. Department of Commerce, Economics and Statistics Administration.

†U.S. Dept. of Commerce, Bureau of the Census. *County and City Data Book {United States] 2000*. Washington, D.C.: U.S. Department of Commerce, Economics and Statistics Administration.

‡U.S. Dept of Commerce, Bureau of the Census. *County and City Data Book [United States] 1956, 1967, 1972, 1977, 1983* (computer files). Washington, D.C.: U.S. Department of Commerce, Bureau of the Census.

§U.S. Dept of Commerce, Bureau of the Census. 1989. *County and City Data Book [United States], 1988* [computer file]. Washington, D.C.: U.S. Department of Commerce, Bureau of the Census.

**U. S. Dept. of Commerce, Bureau of the Census. Table C3. Per Capita Income by County, Income in 1989 CPI-U Adjusted Dollars. In *Censuses of Population and Housing, 1960, 1970, 1980, 1990*. Washington, D.C.: U.S. Department of Commerce, Bureau of the Census.

# CHAPTER 5
# RESULTS AND DISCUSSION

## CANCER MORTALITY ANALYSIS

### Bladder Cancer Mortality

Drinking water arsenic level was not significantly related to bladder cancer mortality (SMRs) in any of the models considered. This was true when deaths for all age groups and decades were combined (1950s, 1960s, 1970s, 1980s, and 1990s) (Table 5.1) or when analysis was restricted to the population 50 years of age and older for all decades combined (Table 5.2). As shown in Table 5.3, the effect of drinking water arsenic on mortality did not achieve even a 20% significance level for either exposure level (drinking water arsenic levels of 11-19 μg/L or 20+ μg/L) for either males or females. Incorporating neighboring county effects, using either definition of a neighboring county as described in the methods did not alter these results.

The most accurate analysis is presented in Table 5.3. This model included county socioeconomic characteristics obtained from U.S. Bureau of the Census data for each decade. Since much of the U.S. Census information available in later decades was not available in the 1950 U.S. Census, we restricted this analysis to deaths occurring between 1960 and 1999. In this model, an additional level of decade was added, so that if county socioeconomic characteristics changed over the 50-year period, these changes were reflected in the data incorporated into the analysis. The decade-specific analysis (1960s, 1970s 1980s and 1990s) was, therefore, summarized across all decades (1960-1999).

Mortality rates vary geographically, and mortality rates in neighboring areas are often strongly related to mortality rates for the study population. In our analyses, the neighboring county mortality had a highly significant relationship with the study area mortality ($p<0.01$ to $p<0.001$) for most analyses. However, in the model shown in Table 5.3, in which county characteristics were measured for each decade, 1960 to 1999, only male bladder cancer mortality was strongly related to neighboring county mortality. Only contiguous counties were used to calculate a neighboring county effect in this model; however, other analyses showed that whether the neighboring county was identified as a contiguous county or by distance from county centroid made very little difference.

A number of covariates were associated with increased or reduced bladder cancer mortality. The coefficients for these covariates are measured in the presence of other covariates and may not reflect the effects of the covariate in the absence of other covariates. Counties designated as a metropolitan area were found to have higher rates of bladder cancer mortality for both male and female populations. This was true in all three analyses: 1) all decades combined analyses for the total population (Table 5.1), 2) people 50 years of age and older (Table 5.2) and 3) analyses adjusted using decade-specific covariates (Table 5.3). Counties with a higher percentage of the population employed in manufacturing had higher rates of male bladder cancer mortality in all three analyses. Conversely, counties with more persons per household were found to have lower bladder cancer mortality rates in all three analyses (Tables 5.1-5.3).

In the analysis using decade-specific covariates, counties with higher percentages of the population with at least an undergraduate college degree had lower bladder cancer death rates. When the analysis was restricted to the population age 50 years and older, this covariate was statistically significant only for females. The same was true for higher levels of per capita personal income; higher per capita personal income was related to a lower bladder cancer mortality rate when decade-specific covariates were used, but the relationship was not observed for males when only a single set of covariates was used for the 50-year period.

**Table 5.1**
**Bladder cancer mortality, males and females, all decades combined**

| | Females | | | Males | | |
|---|---|---|---|---|---|---|
| | No neighboring county adjustment | Adjustment for adjacent county mortality rates (method #1) | Adjustment for nearby county mortality rates (method #2) | No neighboring county adjustment | Adjustment for adjacent county mortality rates (method #1) | Adjustment for nearby county mortality rates (method #2) |
| | Parameter estimate (SE) | | | | | |
| Neighboring county effect | | 0.054 (0.016) * | 0.067 (0.023) * | | 0.087 (0.014) * | 0.11 (0.020) * |
| *Covariates:* | | | | | | |
| *Arsenic level (reference <10 μg/L):* | | | | | | |
| 11-19 μg/L | 0.033 (0.073) | 0.015 (0.075) | 0.0033 (0.081) | 0.0015 (0.062) | -0.0054 (0.058) | 0.030 (0.060) |
| $\geq$ 20 μg/L | -0.096 (0.136) | -0.086 (0.136) | -0.108 (0.137) | 0.011 (0.095) | 0.019 (0.089) | 0.018 (0.092) |
| MSA area | 0.072 (0.028) * | 0.078 (0.029) * | 0.073 (0.029) * | 0.084 (0.024) * | 0.101 (0.023) * | 0.093 (0.022) * |
| % population with at least a 4 year college degree | -0.012 (0.0025) * | -0.0098 (0.0026) * | -0.0098 (0.0027) * | -0.0065 (0.0022) * | -0.0054 (0.0021) * | -0.0051 (0.0020) * |
| Persons per household | -0.246 (0.053) * | -0.249 (0.058) * | -0.238 (0.061) * | -0.278 (0.043) * | -0.329 (0.047) * | -0.311 (0.050) * |
| % county land in farms | -0.00097 (0.00046) * | -0.00086 (0.00053) | -0.00085 (0.00055) | -0.00005 (0.00039) | -0.00009 (0.00046) | -0.00001 (0.00046) |
| Per capita personal income | 0.037 (0.0076) * | 0.027 (0.0084) * | 0.025 (0.0090) * | 0.030 (0.0064) * | 0.012 (0.0067) | 0.013 (0.0068) |
| % employed in manufacturing | | | | 0.0025 (0.00070) * | 0.0016 (0.00075) * | 0.0015 (0.00070) * |

*significant at $p<0.05$

**Table 5.2**
**Bladder cancer mortality, males and females age ≥50 years, all decades combined**

| | Females | | Males | |
|---|---|---|---|---|
| | No neighboring county adjustment | Adjustment for adjacent county mortality rates (method #1) | No neighboring county adjustment | Adjustment for adjacent county mortality rates (method #1) |
| | Parameter estimate (SE) | | | |
| Neighboring county effect | | 0.057 (0.016) * | | 0.092 (0.014) * |
| *Covariates:* | | | | |
| *Arsenic level* | | | | |
| *(reference <10 µg/L):* | | | | |
| 11-19 µg/L | -0.0093 (0.074) | -0.020 (0.078) | 0.012 (0.062) | -0.0094 (0.059) |
| ≥ 20 µg/L | -0.071 (0.135) | -0.059 (0.135) | 0.011 (0.099) | 0.0066 (0.090) |
| MSA area | 0.062 (0.028) * | 0.069 (0.029) * | 0.12 (0.024) * | 0.11 (0.022) * |
| % population with at least a 4 year college degree | -0.010 (0.0025) * | -0.0089 (0.0026) * | -0.0028 (0.0022) | -0.0036 (0.0021) |
| *Persons per household* | -0.25 (0.054) * | -0.25 (0.059) * | -0.36 (0.041) * | -0.36 (0.046) * |
| % county land in farms | -0.00067 (0.00046) | -0.00064 (0.00055) | -0.00024 (0.00039) | -0.00025 (0.00047) |
| Per capita personal income | 0.035 (0.0078) * | 0.024 (0.0085) * | -0.00018 (0.00044) | -0.00016 (0.00042) |
| % employed in manufacturing | | | 0.0031 (0.00072) * | 0.0020 (0.00076) * |

*significant at $p<0.05$

**Table 5.3**
**Bladder cancer mortality in males and females by individual decades**

| | Females | | Males | |
|---|---|---|---|---|
| | No neighboring county adjustment | Adjustment for adjacent county mortality rates (method #1) | No neighboring county adjustment | Adjustment for adjacent county mortality rates (method #1) |
| | Parameter estimate (SE) | | | |
| Neighboring county effect | | 0.0033 (0.0025) | | 0.075 (0.014) * |
| *Covariates:* | | | | |
| *Arsenic level (reference <10 μg/L):* | | | | |
| 11-19 μg/L | 0.0033 (0.070) | 0.0030 (0.072) | -0.021 (0.061) | -0.028 (0.061) |
| ≥ 20 μg/L | -0.163 (0.146) | -0.158 (0.146) | 0.087 (0.097) | 0.079 (0.092) |
| MSA area | 0.095 (0.027) * | 0.096 (0.027) * | 0.087 (0.022) * | 0.093 (0.022) * |
| % population with at least a 4 year college degree | -0.013 (0.0024) * | -0.013 (0.0024) * | -0.011 (0.0021) * | -0.008 (0.0021) * |
| *Persons per household* | -0.247 (0.047) * | -0.249 (0.048) * | -0.244 (0.037) * | -0.249 (0.040) * |
| % county land in farms | -0.00073 (0.00044) | -0.00072 (0.00045) | -0.00010 (0.00037) | -0.00035 (0.00042) |
| Per capita personal income | 0.028 (0.0064) * | 0.028 (0.0064) * | 0.021 (0.0050) * | 0.012 (0.0053) * |
| % employed in manufacturing | | | 0.0035 (0.00079) * | 0.0026 (0.00084) * |

*significant at $p<0.05$

## Lung Cancer Mortality

The highest level of arsenic in drinking water, ≥ 20 μg/L was significantly associated with higher lung cancer SMRs in female populations when deaths in all age groups during all decades were combined and county population characteristics measured the 1980 U.S. Census were related to the SMRs in the regression model ($p<0.05$). The same finding was apparent when only deaths occurring to women age 50 years and older were considered ($p<0.05$). However, statistical significance disappeared when female mortality rates for all ages were adjusted for adjacent or nearby county female lung cancer mortality rates. The statistically significant elevation in female lung cancer SMRs was also seen in deaths to women age 50 years and older and this relationship persisted even after adjustment for mortality rates in adjacent counties (Table 5.5). However, when the analysis was conducted using Census data for each decade, the adjustment eliminated any statistically significant differences (Table 5.6). In conclusion, the analysis of female lung cancer deaths from 1960 to 1999, adjusting for population characteristics measured in each decade and adjusted for adjacent county mortality rates did not show evidence of elevated lung cancer mortality rates.

For other characteristics, the findings of the lung cancer analyses were similar to the findings of the bladder cancer analyses (Tables 5.4, 5.5,and 5.6). In all lung cancer mortality analyses, counties designated as metropolitan areas were found to have higher mortality rates; while counties with higher percentages of the population with at least an undergraduate degree and counties with more persons per household were found to have lower lung cancer mortality rates. Counties with higher per capita personal income were found to have higher female, but not male, lung cancer mortality rates. Having a higher percentage of the population employed in manufacturing was found to be significantly associated with higher mortality among male populations in some but not all of the analyses. In general, counties having higher percentages of county acreage in farming had lower rates of male lung cancer mortality among in all analyses, but this factor was not consistently related to female lung cancer mortality.

## Effect of Neighboring County Adjustment

Total calculated variation (Langford et al. 1999) in mortality due to differences between counties and to differences between adjacent areas for bladder cancer and lung cancer are shown in Tables 5.7 and 5.8. Similar results were achieved using the second definition of nearby counties based on distance between the county centroids in the model in which we combined deaths for all decades and all ages. While there is still a large amount of variance unexplained by the models, there is strong support for the proposition that neighboring counties have similar mortality rates for lung and bladder cancer. There is evidence of significant variation between counties and between neighboring areas (Table 5.8). The evidence was less consistent in bladder cancer mortality analyses. In the older population analysis of all decades combined, county and neighboring area variation was significant, but variation between counties was not found to be significant, nor was there significant variation between counties and neighboring areas with respect to bladder cancer mortality among females in the individual decades analysis.

**Table 5.4**
**Lung cancer mortality, males and females, all decades combined**

| | Females | | | Males | | |
|---|---|---|---|---|---|---|
| | No neighboring county adjustment | Adjustment for adjacent county mortality rates (method #1) | Adjustment for nearby county mortality rates (method #2) | No neighboring county adjustment | Adjustment for adjacent county mortality rates (method #1) | Adjustment for nearby county mortality rates (method #2) |
| | Parameter estimate (SE) | | | | | |
| Neighboring county effect | | 0.12 (0.015) * | 0.16 (0.023) * | | 0.100 (0.011) * | 0.165 (0.023) * |
| *Covariates:* | | | | | | |
| *Arsenic level (reference <10 μg/L):* | | | | | | |
| 11-19 μg/L | 0.021 (0.055) | 0.022 (0.047) | 0.055 (0.049) | -0.059 (0.046) | -0.0084 (0.036) | 0.059 (0.049) |
| ≥ 20 μg/L | 0.15 (0.070) * | 0.09 (0.060) | 0.11 (0.062) | -0.027 (0.059) | 0.015 (0.046) | 0.109 (0.062) |
| MSA area | 0.19 (0.027) * | 0.14 (0.024) * | 0.13 (0.024) * | 0.155 (0.022) * | 0.104 (0.019) * | 0.133 (0.024) * |
| % population with at least a 4 year college degree | -0.020 (0.002) * | -0.015 (0.0019) * | -0.013 (0.0019) * | -0.024 (0.0017) * | -0.018 (0.0015) * | -0.012 (0.0019) * |
| Persons per household | -0.40 (0.036) * | -0.32 (0.040) * | -0.30 (0.044) * | -0.486 (0.030) * | -0.364 (0.032) * | -0.309 (0.044) * |
| % county land in farms | -0.0023 (0.0004) * | -0.0018 (0.0004) * | -0.0019 (0.0004) * | -0.00057 (0.00029) | -0.00021 (0.00035) | -0.0019 (0.00046) * |
| Per capita personal income | 0.022 (0.0049) * | 0.018 (0.0045) * | 0.014 (0.0046) * | 0.0019 (0.0041) | 0.0029 (0.0036) | 0.014 (0.0046) * |
| % employed in manufacturing | | | | 0.0044 (0.00069) * | 0.0028 (0.00063) * | 0.00091 (0.00054) |

*significant at $p<0.05$

**Table 5.5**
**Lung cancer mortality, males and females ≥ 50 years, all decades combined**

| | Females | | Males | |
|---|---|---|---|---|
| | No neighboring county adjustment | Adjustment for adjacent county mortality rates (method #1) | No neighboring county adjustment | Adjustment for adjacent county mortality rates (method #1) |
| | Parameter estimate (SE) | | | |
| Neighboring county effect | | 0.16 (0.020) * | | 0.10 (0.011) * |
| *Covariates:* | | | | |
| *Arsenic level (reference <10 µg/L):* | | | | |
| 11-19 µg/L | 0.085 (0.062) | 0.075 (0.051) | -0.053 (0.047) | -0.0064 (0.037) |
| ≥ 20 µg/L | 0.23 (0.077) * | 0.14 (0.066) * | -0.026 (0.060) | 0.025 (0.048) |
| MSA area | 0.26 (0.030) * | 0.20 (0.027) * | 0.21 (0.023) * | 0.15 (0.019) * |
| % population with at least a 4 year college degree | -0.017 (0.0022) * | -0.012 (0.0020) * | -0.022 (0.0017) * | -0.016 (0.0015) * |
| *Persons per household* | -0.31 (0.041) * | -0.26 (0.045) * | -0.42 (0.031) * | -0.31 (0.033) * |
| % county land in farms | -0.0039 (0.00039) * | -0.0030 (0.00048) * | -0.0013 (0.00030) * | -0.00076 (0.00036) * |
| Per capita personal income | 0.025 (0.0054) * | 0.018 (0.0050) * | 0.0075 (0.0041) | 0.0060 (0.0036) |
| % employed in manufacturing | | | 0.0021 (0.00051) * | 0.0010 (0.00045) * |

*significant at $p<0.05$

**Table 5.6**
**Lung cancer mortality, males and females, by individual decades**

| | Females | | Males | |
|---|---|---|---|---|
| | No neighboring county adjustment | Adjustment for adjacent county mortality rates (method #1) | No neighboring county adjustment | Adjustment for adjacent county mortality rates (method #1) |
| | Parameter estimate (SE) | | | |
| Neighboring county effect | | 0.186 (0.020) * | | 0.169 (0.016) * |
| *Covariates:* | | | | |
| *Arsenic level (reference <10 µg/L):* | | | | |
| 11-19 µg/L | -0.0032 (0.062) | 0.046 (0.048) | -0.108 (0.062) | 0.015 (0.038) |
| ≥ 20 µg/L | 0.117 (0.084) | 0.078 (0.065) | -0.062 (0.077) | 0.018 (0.053) |
| MSA area | 0.224 (0.026) * | 0.159 (0.021) * | 0.147 (0.025) * | 0.107 (0.017) * |
| % population with at least a 4 year college degree | -0.019 (0.0018) * | -0.013 (0.0017) * | -0.016 (0.0014) * | -0.013 (0.0012) * |
| *Persons per household* | -0.330 (0.028) * | -0.287 (0.029) * | -0.289 (0.020) * | -0.245 (0.019) * |
| % county land in farms | -0.00080 (0.00032) * | -0.00023 (0.00035) | -0.0011 (0.00026) * | 0.00092 (0.00026) * |
| Per capita personal income | 0.019 (0.0031) * | 0.014 (0.0032) * | 0.0029 (0.0023) | 0.00032 (0.0022) |
| % employed in manufacturing | | | 0.0014 (0.00040) * | 0.00057 (0.00038) |

*significant at $p<0.05$

**Table 5.7**
**Bladder cancer mortality, county and neighboring area variation**

| Analysis approach | Female: Total variation* | Female: % mortality due to neighboring area effects† | Male: Total variation* | Male: % due to neighboring area effects† |
|---|---|---|---|---|
| All decades combined | 0.0084 | 80% | 0.014 | 91% |
| All decades combined, population age ≥ 50 years | 0.0089 | 83% | 0.014 | 91% |
| Individual decades | 0.00 | – | 0.012 | 92% |

*Calculated as the sum of county variance and (neighborhood variance/mean # neighbors)
†Effects due to neighboring areas defined as contiguous counties

**Table 5.8**
**Lung cancer mortality, county and neighboring area variation**

| Analysis approach | Female: Total variation* | Female: % mortality due to neighboring area effects† | Male: Total variation* | Male: % due to neighboring area effects† |
|---|---|---|---|---|
| All decades combined | 0.047 | 81% | .016 | 52% |
| All decades combined, population age ≥ 50 years | 0.025 | 45% | .031 | 50% |
| Individual decades | 0.030 | 82% | .030 | 86% |

*Calculated as the sum of county variance and (neighborhood variance/mean # neighbors)
†Effects due to neighboring areas defined as contiguous counties

The magnitude of total variation and the amount of explained variation due to county and neighborhood effects is greater for lung cancer mortality than bladder cancer mortality. This can also be seen graphically in maps of the county SMRs ranked according to quintiles (Figures 5.1-5.9). Clusters (areas where counties have similarly high or low SMRs) are much more evident in maps of lung cancer mortality SMRs than in maps of bladder cancer SMRs. As an illustration of the effect of rare occurrences on standard SMR calculation, Figure 5.4 is included as an alternative to Figure 5.3. To diminish the variation in SMR due to small denominator calculations, SMRs in Figure 5.4 have been calculated after adding one (1) to both the observed and expected bladder cancer deaths. This serves to move some counties to a higher or lower quintile.

## CANCER INCIDENCE

As outlined in Chapter 3, the relationship between arsenic exposure and cancer incidence was examined for New Mexico and Utah, the only two study states with Surveillance, Epidemiology and End Results (SEER) cancer registries, for 1973-1999. It should be noted that in this analysis, there were no counties with mean drinking water arsenic levels exceeding 20 μg/L. While results are reported here, it must be noted that results pertain only to bladder and lung cancer incidence in Utah and New Mexico, with only two large metropolitan areas represented, Bernalillo county and Salt Lake county.

A higher level of arsenic in drinking water was not significantly related to higher bladder or lung cancer incidence. Since only two states are represented in the analysis, there are limitations in the factors that can be adequately tested. Because of the strong influence of the Mormon Church in Utah, we added a variable to identify cases from Utah. This is especially important since both bladder and lung cancer are smoking-related diseases and the smoking prevalence in Utah is very low. We used the same U.S. Bureau of the Census county socioeconomic covariates in this analysis as we did in the mortality analysis, adjusted for the appropriate decade. Relationships between bladder cancer incidence and covariates were similar to those for bladder cancer mortality (Table 5.9). None of the socioeconomic variables were statistically significantly related to incidence (SIR) of bladder cancer.

With respect to the analysis of lung cancer incidence in females, counties designated as metropolitan area were not found to have higher lung cancer incidence, while Utah residence, counties with a higher percentage of the population having at least an undergraduate education, and counties with a higher percentage of acreage in farmland had lower SIRs (Table 5.10). As in the mortality analysis, counties with higher levels of per capita income had higher lung cancer SIRs.

For male lung cancer incidence the results were slightly different. Neither residence in Utah nor having a higher fraction of the population with a college degree were related to a higher lung cancer SIR. A higher fraction of the county land was in farms and more people per household were associated with lower male lung cancer SIR after adjusting for neighboring county SIRs. Conversely, the percent of the population employed in manufacturing and per capita personal income were not related to lung cancer SIR (Table 5.10).

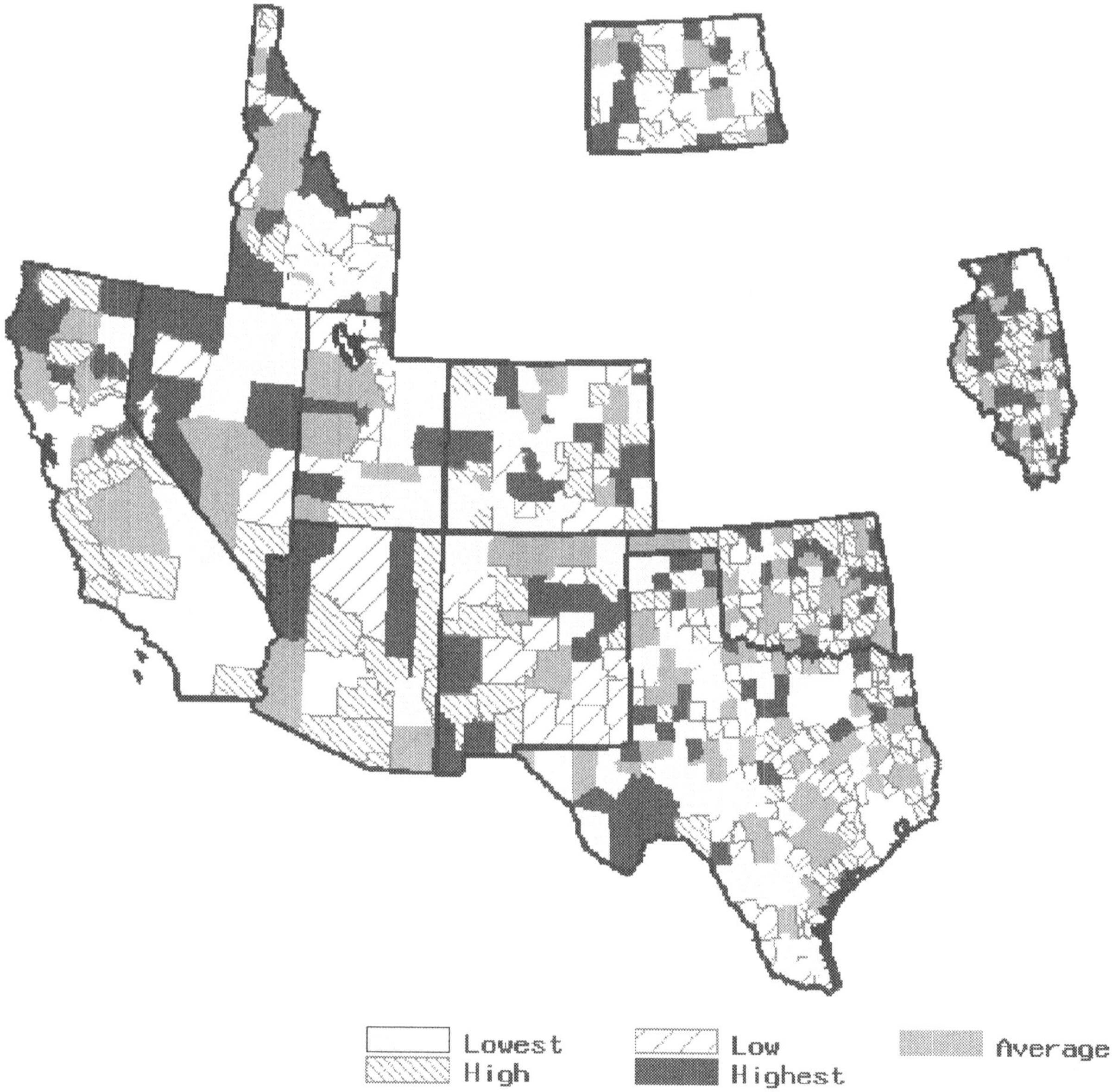


**Figure 5.1 Bladder cancer SMRs in females, 1950-1999**

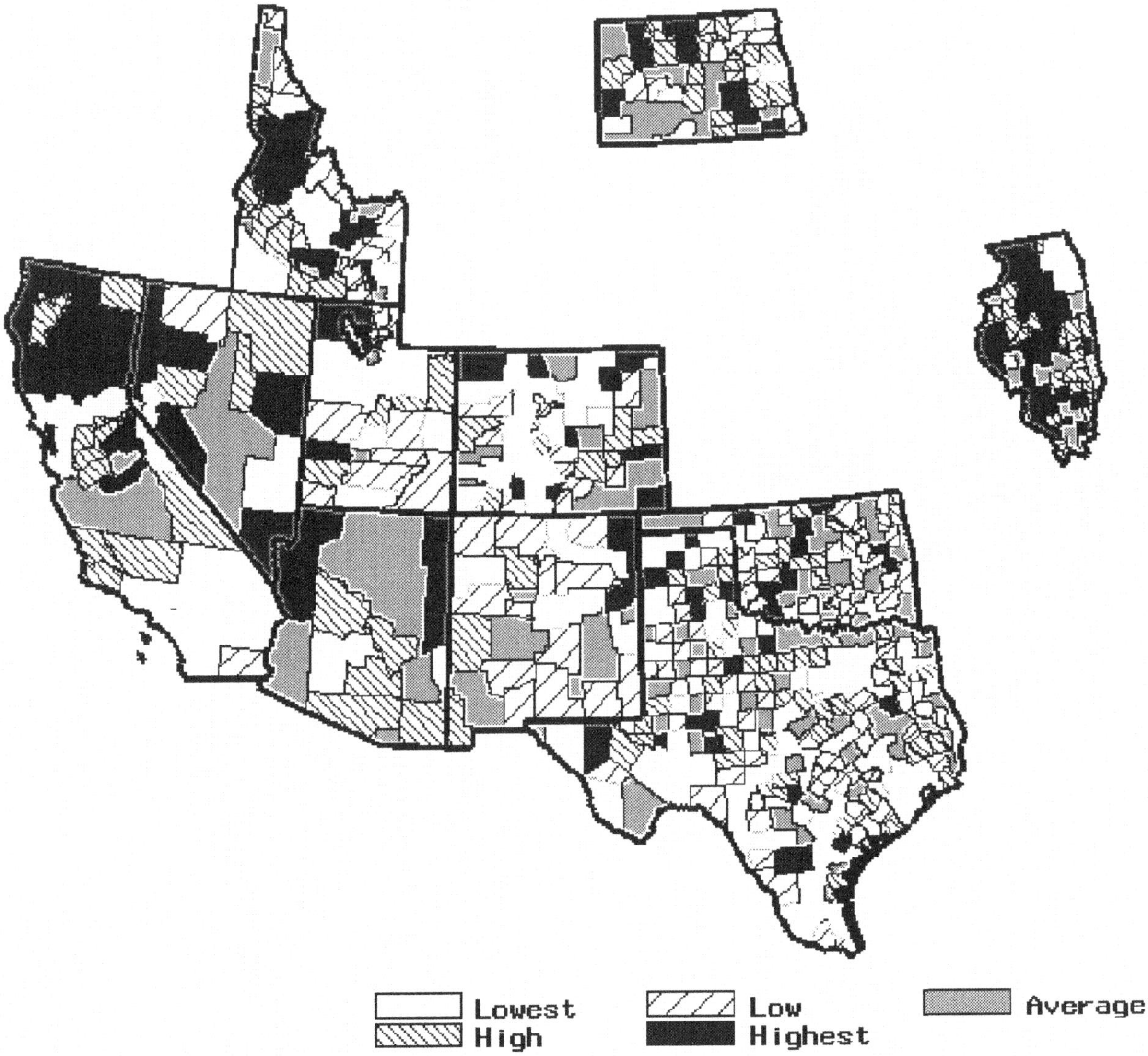


Figure 5.2 Bladder cancer SMRs in males, 1950-1999

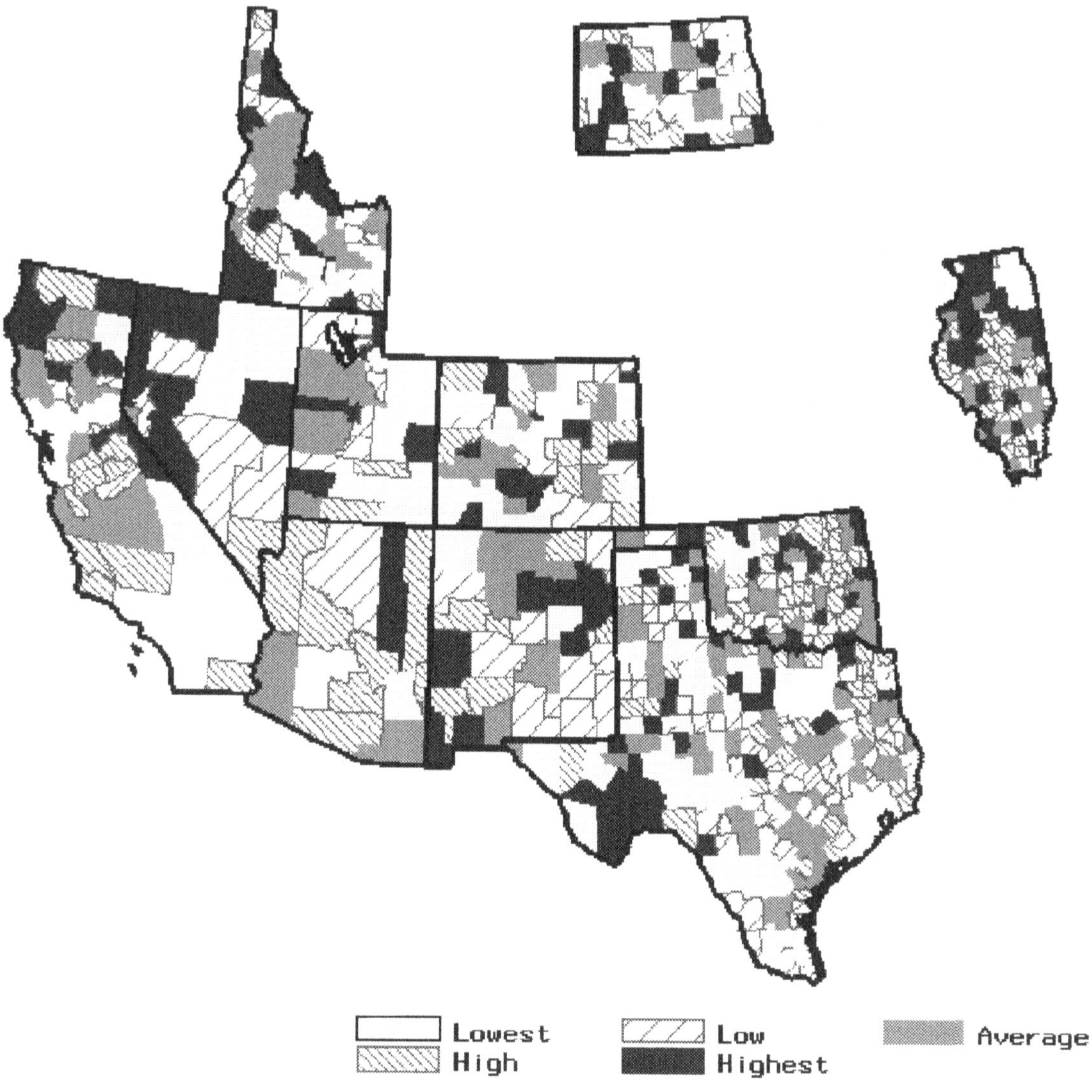


**Figure 5.3 Bladder cancer SMRs 1950-1999 in females age ≥ 50 years**

**Figure 5.4 Adjusted bladder cancer SMRs 1950-1999 in females (observed and expected values increased by 1 prior to calculating SMR)**

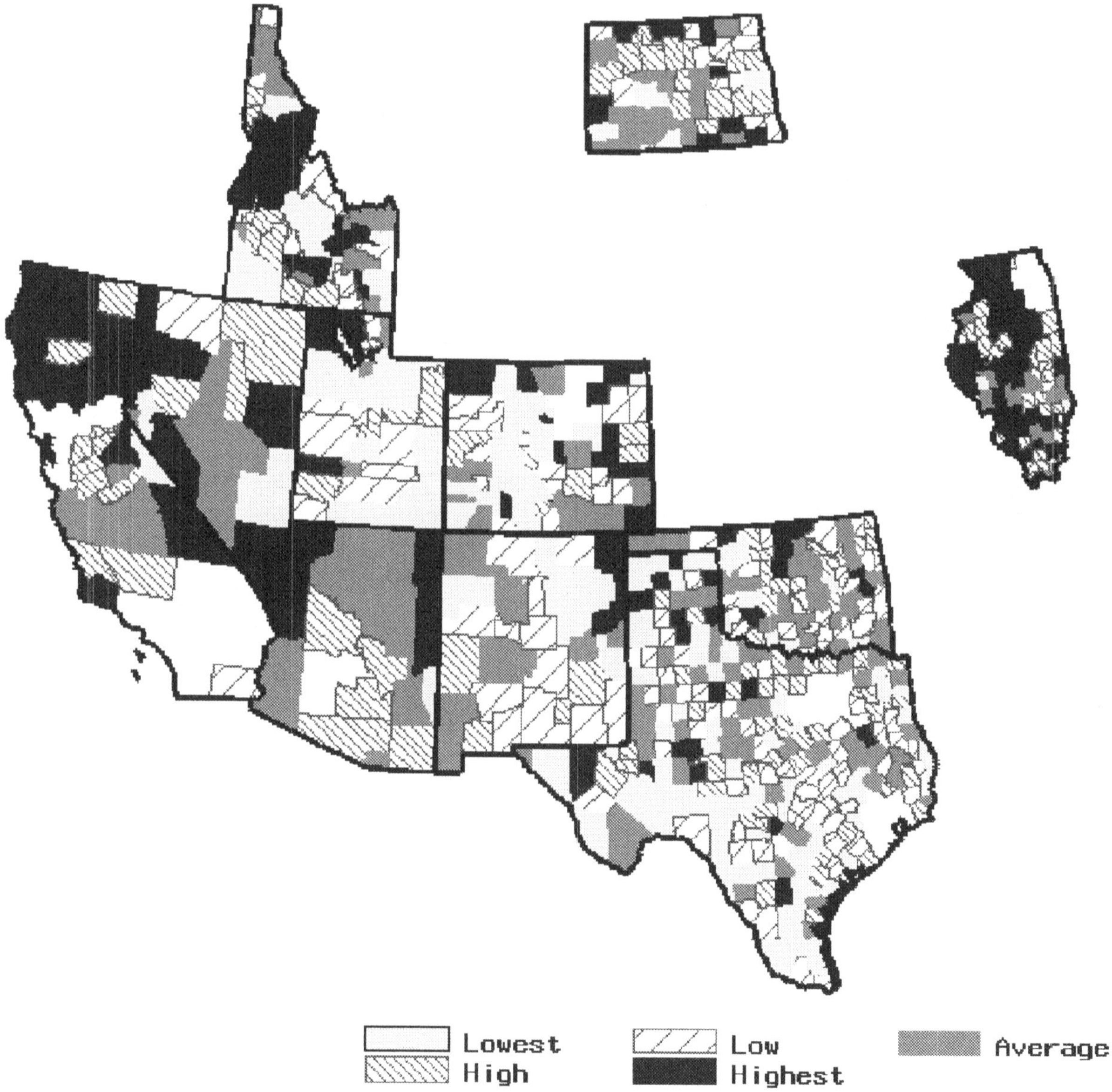


**Figure 5.5 Bladder cancer SMRs 1950-1999 in males age ≥ 50 years**

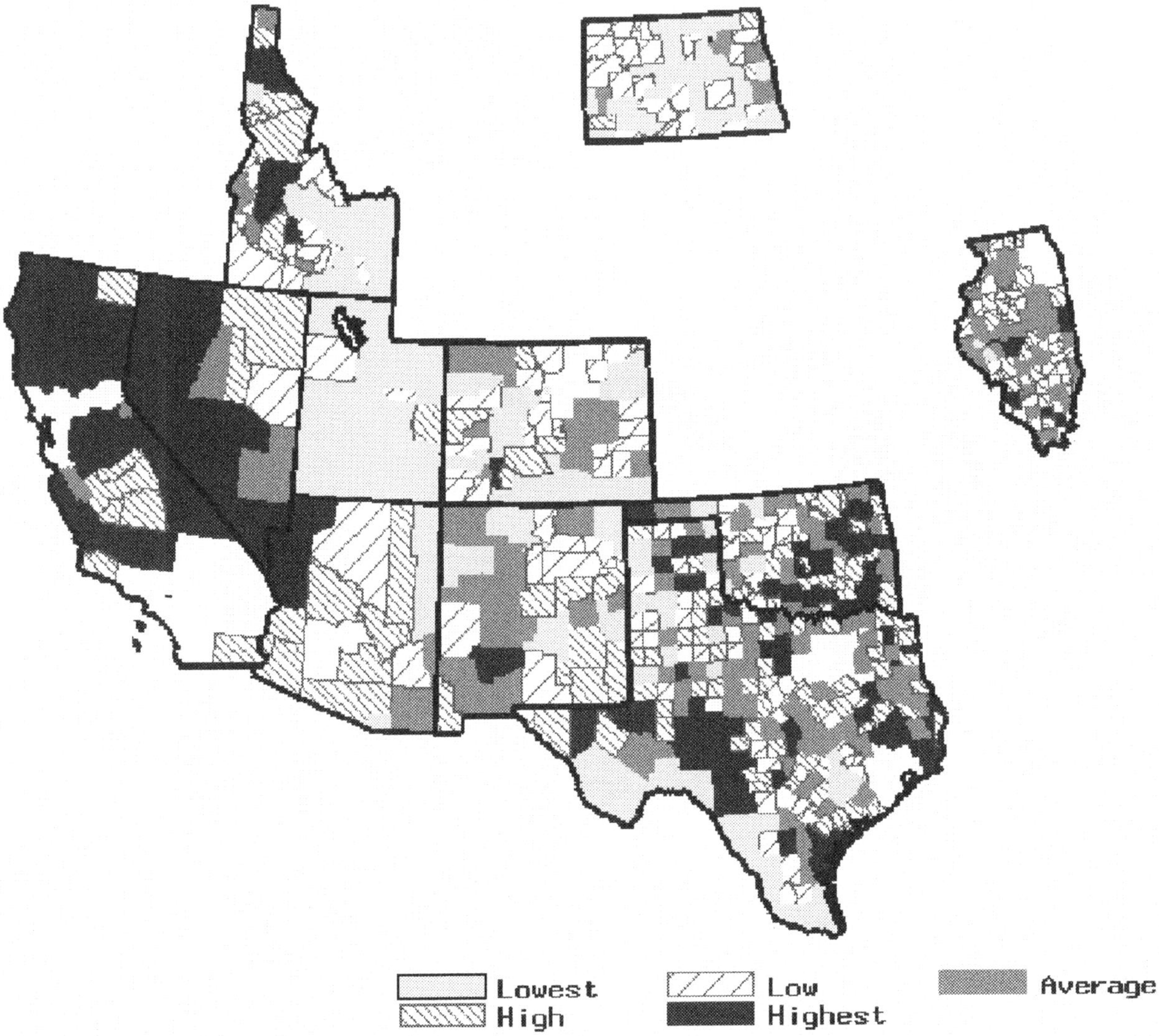


**Figure 5.6 Lung cancer SMRs 1950-1999 in females**

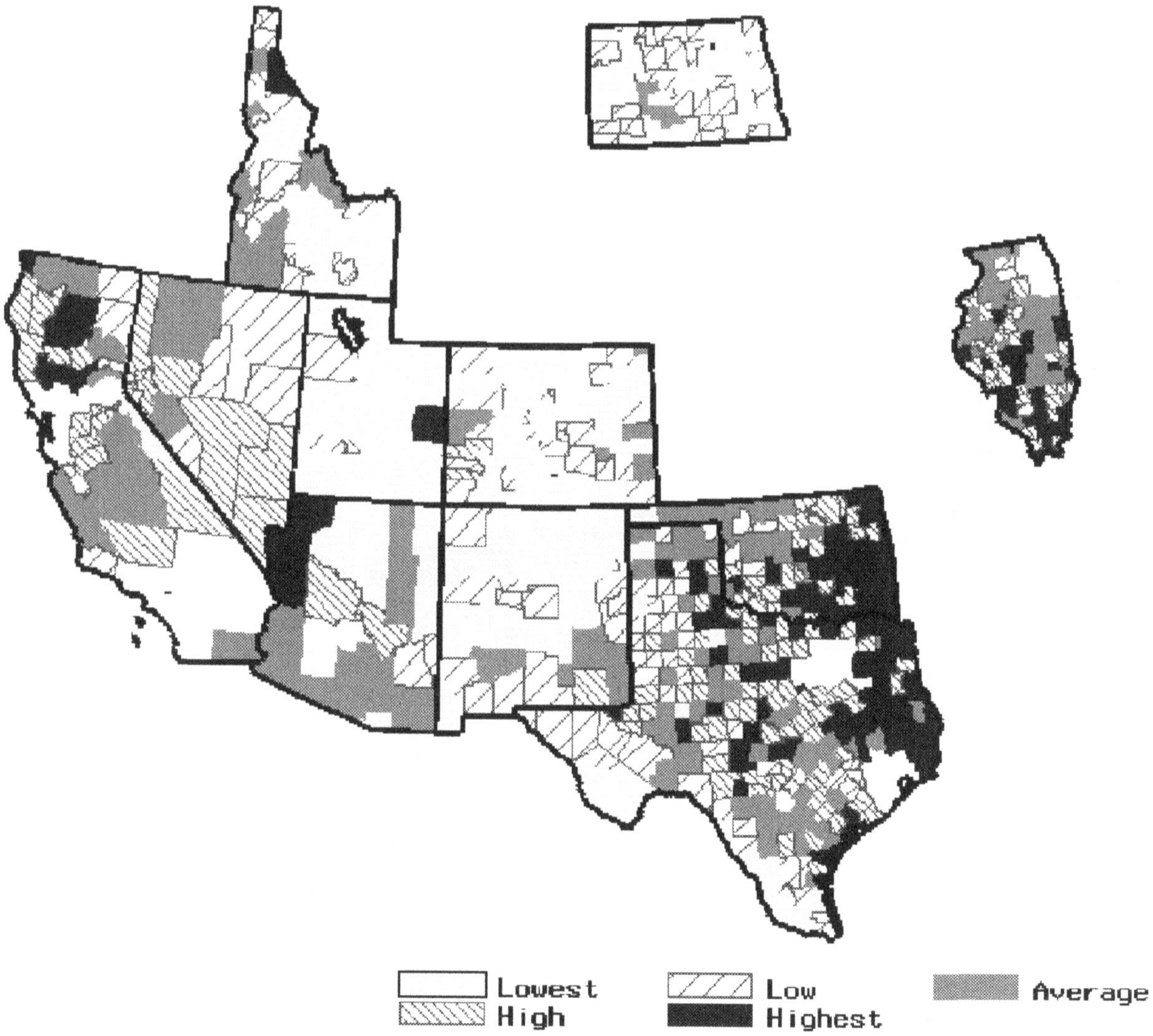


Figure 5.7 Lung cancer SMRs 1950-1999 in males

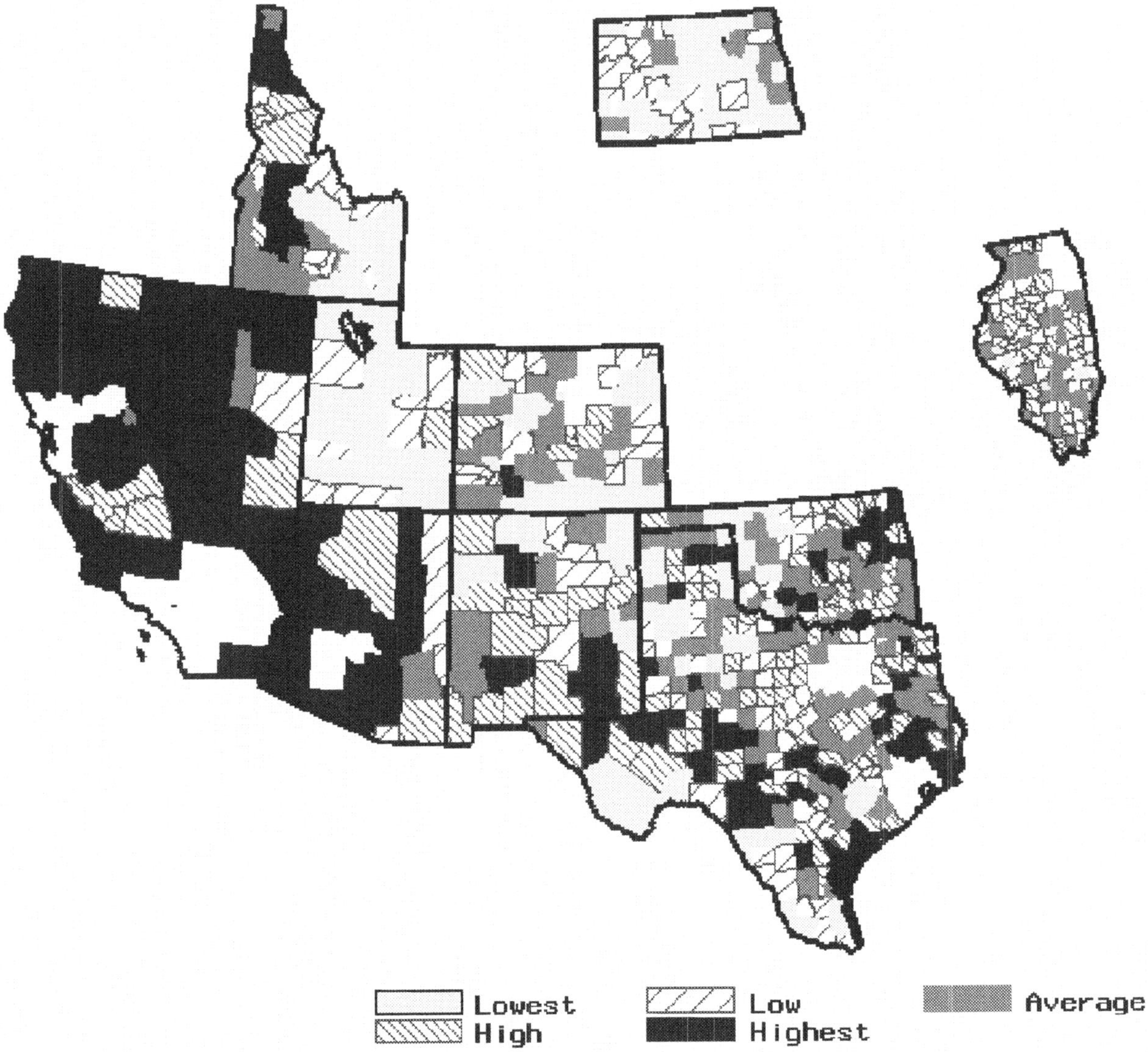


**Figure 5.8 Lung cancer SMRs 1950-1999 in females age ≥ 50 years**

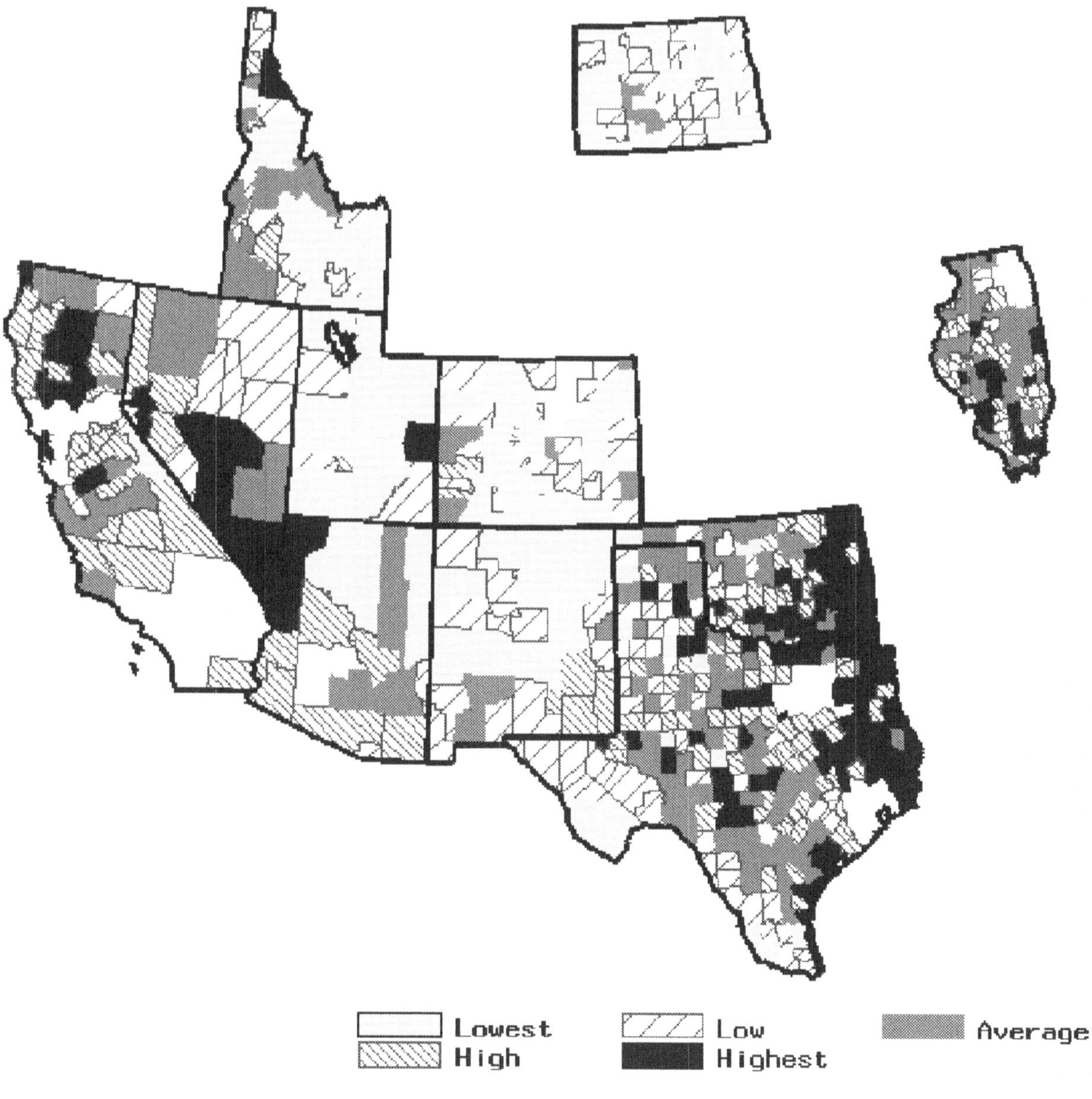


**Figure 5.9 Lung cancer SMRs 1950-1999 in males age ≥ 50 years**

**Table 5.9**
**Bladder cancer incidence 1973-1999, males and females**

| | Females | | | Males | | |
|---|---|---|---|---|---|---|
| | Adjustment for time period (decade) | Adjustment for differences between counties | Adjustment for adjacent county mortality rates (#1) | Adjustment for time period (decade) | Adjustment for differences between counties | Adjustment for adjacent county mortality rates (#1) |
| | Parameter Estimate (SE) | | | | | |
| Variance between counties | | 0.011 (0.0075) | 0.0086 (0.0073) | | 0.011 (0.0063) | 0.0072 (0.0059) |
| Neighboring county effect | | | 0.018 (0.025) | | | 0.0059 (0.0054) |
| *Covariates:* | | | | | | |
| *Arsenic level >10 μg/L* | 0.125 (0.069) | 0.200 (0.107) | 0.211 (0.114) | 0.077 (0.042) | 0.096 (0.081) | 0.083 (0.084) |
| MSA area | 0.097 (0.062) | 0.108 (0.080) | 0.109 (0.082) | 0.067 (0.034) | 0.101 (0.061) | 0.100 (0.060) |
| County in Utah | -0.063 (0.070) | -0.087 (0.090) | -0.098 (0.101) | 0.052 (0.042) | 0.052 (0.063) | -0.052 (0.070) |
| % population with at least a 4 year college degree | -0.0017 (0.0053) | -0.0019 (0.0063) | -0.00097 (0.0066) | -0.0093 (0.0030) | -0.0065 (0.0045) | -0.0060 (0.0042) |
| Persons per household | -0.156 (0.111) | -0.176 (0.134) | -0.173 (0.136) | -0.113 (0.062) | -0.085 (0.084) | -0.068 (0.092) |
| % county land in farms | -0.00022 (0.0010) | -0.00014 (0.0011) | -0.00007 (0.0012) | -0.00064 (0.00059) | -0.00060 (0.00071) | -0.00072 (0.00072) |
| Per capita personal income | 0.0022 (0.013) | 0.0044 (0.014) | 0.0055 (0.014) | 0.016 (0.0071) | 0.0059 (0.0096) | 0.0040 (0.094) |
| % employed in manufacturing | | | | 0.0031 (0.0015) | -0.0021 (0.0019) | -0.0020 (0.0020) |

*significant at $p<0.05$

**Table 5.10**
**Lung cancer incidence 1973-1999, males and females**

| | Females | | | Males | | |
|---|---|---|---|---|---|---|
| | Adjustment for time period (decade) | Adjustment for differences between counties | Adjustment for adjacent county mortality rates (#1) | Adjustment for time period (decade) | Adjustment for differences between counties | Adjustment for adjacent county mortality rates (#1) |
| | Parameter Estimate (SE) | | | | | |
| Variance between time periods (decade) | 0.173 (0.019) * | 0.163 (0.020) * | 0.162 (0.019) * | 0.138 (0.015) * | 0.062 (0.0084) * | 0.062 (0.0084) * |
| Variance between counties | | 0.013 (0.012) | 0.011 (0.010) | | 0.088 (0.023) * | 0.053 (0.028) |
| Neighboring county effect | | | 0.026 (0.039) | | | 0.193 (0.154) |
| *Covariates:* | | | | | | |
| *Arsenic level >10 μg/L* | -0.086 (0.131) | -0.084 (0.146) | -0.085 (0.146) | 0.076 (0.119) | 0.082 (0.179) | 0.079 (0.168) |
| MSA area | 0.175 (0.096) | 0.180 (0.102) | 0.180 (0.104) | 0.062 (0.085) | 0.038 (0.124) | 0.044 (0.121) |
| County in Utah | -0.640 (0.097) * | -0.635 (0.102) * | -0.625 (0.114) * | -0.221 (0.088) * | -0.127 (0.118) | -0.062 (0.161) |
| % population with at least a 4 year college degree | -0.029 (0.0067)* | -0.029 (0.0071) * | -0.029 (0.0072)* | -0.027 (0.0060)* | -0.014 (0.0079) | -0.013 (0.0078) |
| Persons per household | -0.192 (0.125) | -0.193 (0.134) | -0.198 (0.133) | -0.104 (0.112) | -0.286 (0.135) * | -0.294 (0.141) * |
| % county land in farms | -0.0047 (0.0014)* | -0.0046 (0.0015)* | -0.0043 (0.0016)* | -0.0016 (0.0012) | -0.00048 (0.0018)* | -0.00088 (0.0020)* |
| Per capita personal income | 0.032 (0.016) * | 0.031 (0.016) * | 0.030 (0.016) * | 0.033 (0.014) * | 0.010 (0.014) | 0.0097 (0.014) |
| % employed in manufacturing | | | | -0.00086 (0.0021) | -0.0010 (0.0022) | -0.0011 (0.0022) |

*significant at $p<0.05$

# CHAPTER 6
# SUMMARY AND CONCLUSIONS

## SUMMARY OF FINDINGS

Higher levels of arsenic in drinking water were not associated with higher rates of bladder or lung cancer mortality when statistical models were adjusted for neighboring county mortality rates. Neighboring county mortality rates were strongly associated with county male and female lung cancer death rates, but less so for bladder cancer death rates. In fact, only county male bladder cancer mortality rates were significantly related to neighboring county mortality rates.

The individual decade analyses of lung and bladder cancer mortality are the most accurately adjusted estimates of effect, since the socioeconomic covariates for each decade were derived from the U.S. Bureau of the Census data for that decade. In these analyses, the neighboring county adjustment controlled 92% of the county-level variance in male bladder death rates, 82% of the county-level variance in female lung cancer death rates, and 86% of the county-level variance in male lung cancer death rates.

Being in a metropolitan statistical area (MSA) was strongly related to an increased risk of bladder and lung cancer death. Having a larger mean household size and having a higher fraction of the population with at least a 4-year college degree were strongly related to a reduced risk of bladder and lung cancer death. Other factors were less consistently related to mortality rates.

These relationships were different for cancer incidence in New Mexico and Utah. This may be due to the unique characteristics of the Utah population and the low rates of these diseases in New Mexico. Residing in a metropolitan statistical area was not statistically associated with an increased risk of either bladder or lung cancer. The effect of neighboring county cancer incidence was considered, but due to the small numbers of counties in each state (32 in New Mexico; 29 in Utah), its effect was muted. The effect of neighboring county incidence was only predictive for lung cancer incidence in males, and this relationship was marginal ($p<0.05$).

In most of the country, rural areas have a lower death rate from lung and bladder cancer for both males and females. We observed this relationship in the mortality analysis for the entire study area.

If drinking water arsenic exposures only marginally increase cancer risks for smokers after long latencies, as suggested by Bates et al. (2003) and Steinmaus et al. (2003), then this study would not have had the power to detect the elevated risks.

## CONCLUSIONS AND LIMITATIONS

This study did not find evidence of increased risk for lung or bladder cancer mortality or incidence from exposure to arsenic in drinking water. The findings are consistent with other recent studies of low dose arsenic health effects and are inconsistent with the NRC predictions of an increase in cancer risk from low dose arsenic exposure (NRC 2001).

It is possible that elevated risks were present but not apparent in the analysis we conducted. There are several reasons an effect could have been missed. First, an analysis of bladder cancer mortality is limited by the fact that many people with bladder cancer do not die from bladder cancer. Secondly, if drinking water arsenic exposure only marginally increases cancer risks for smokers, as suggested by Bates et al (2003) and Steinmaus et al (2003), the magnitude of the elevated public health risk would not have been large enough to detect in our analysis. It should be noted, however, that the NRC report (2001) did not conclude that arsenic-associated cancer risks were restricted to smokers with a smoking history of 40 or more pack-years.

Since the latency between completion of exposure to arsenic-contaminated drinking water and the occurrence of deaths from cancer is relatively long and of unknown length, some health outcomes may not have been captured, as people may have moved away or died from other causes. Alternatively, failure to observe the expected cancers could have been caused by problems with the underlying hypotheses.

As stated in Chapter 2, this is an ecological study and as such, it is subject to the limitations of ecological studies. Since the study relates exposures and outcomes in groups of people, it is possible that the true relationship between the exposure (i.e., arsenic in drinking water) and the outcomes (bladder and lung cancer incidence and mortality) for individuals is very different than that observed for groups of individuals. Secondly, although efforts were made to adjust for confounding factors (i.e., characteristics of the population related to both the risk of disease and exposure to arsenic), it is possible that unadjusted confounding factors could limit our ability to detect health effects from waterborne arsenic.

Although we have attempted to estimate arsenic exposure from drinking water, we cannot determine the arsenic exposure from other sources, such as food. In fact, certain foods, such as grapes and rice, can be a significant source of arsenic exposure. Food exposures could significantly reduce differences in arsenic exposures between populations. This would reduce the effectiveness of any study to detect arsenic-related health risks since it may be difficult to find populations with sufficiently large differences in arsenic exposure.

If an arsenic-related health risk only occurred at the highest arsenic exposure levels, then this study may have limited ability to detect health effects. This would occur because the population exposed to arsenic levels above 20 ppb or 50 ppb is relatively small.

## IMPLICATIONS OF THE FINDINGS

In 2001, the U.S. EPA adopted a new maximum contaminant level (MCL) for drinking water arsenic. During the 2001 NRC deliberations, the author received requests from a member of the NRC panel for data on lung and bladder cancer incidence and mortality for U.S. arsenic-exposed populations. These data were not available at that time because there were no studies evaluating whether U.S. residents exposed to elevated drinking water arsenic levels have elevated cancer risks. This study was a rigorous effort to determine if arsenic-exposed U.S. populations have elevated risks of developing or dying from arsenic-related cancers.

According to the EPA, lowering the arsenic MCL will prevent the occurrence of and death from arsenic-related bladder and lung cancers and possibly cardiovascular and hypertensive diseases (EPA 2000). Judged on the cost per year-of-life-gained, the new MCL of 10 μg/L is among the most expensive public health interventions ever adopted by the U.S. government. The EPA estimated that lowering the arsenic MCL to 10 μg/L will prevent between 6.9 bladder cancer and 33 lung cancer deaths each year in the United States.

The study reported here is important because it provides further evidence that additional studies are needed to evaluate the scientific justification for the new arsenic MCL. As outlined in Chapter 1, the findings of several new arsenic health effects studies are inconsistent with the EPA's predicted elevated cancer risks (Bates et al. 2003, Bates et al. 1995, Buchet and Lison 1998, Lewis et al. 1999, Steinmaus et al. 2003).

## FUTURE RESEARCH

This study made use of a particularly appropriate statistical method (hierarchical modeling) for analyzing ecological data, such as mortality or cancer incidence data, for geographic entities. This statistical modeling approach can be readily applied to a wide range of drinking water-related health issues. Drinking water exposures are much more geographically limited than are exposures that occur through other media, such as air or food. Drinking water is seldom exported beyond the boundary of the service area and most of the drinking water consumed in the service area is supplied by the water utilities in the area. The multi-level hierarchical model used in this study was specifically developed for the types of studies commonly conducted to detect drinking water health effects and may be the optimal approach for such studies. The role of drinking water minerals on risk of sudden cardiac death, for example, can be better addressed with these multi-level hierarchical models than with a traditional ecological study design. Whenever there is either widespread contamination of drinking water that may either increase or reduce health risks, hierarchical modeling is a useful tool for assessing the effects. This statistical method can improve the power of ecological studies by effectively adjusting for some confounding factors.

Studies of cardiovascular risk and drinking water arsenic could be accomplished using a similar design. Multi-level hierarchical models could be used to determine if areas with elevated drinking water arsenic have elevated rates of cardiovascular disease, adjusting for rates of cardiovascular disease in neighboring counties. A prior study by Engel and Smith (1994) suggested that geographic areas with elevated drinking water arsenic have elevated rates of cardiovascular disease. This study was flawed, however, by incorrect assignment of exposed and unexposed population and did not use hierarchical models

Researchers at the University of California are planning to conduct a case-control study of lung cancer in arsenic-exposed areas of Nevada and California. This study will provide additional information on whether excess risk of lung cancer can be observed in U.S. populations exposed to drinking water arsenic. The next step in better understanding the dose-response relationship between arsenic and cancer will probably involve studies of markers of cellular changes that may predict either an increased or reduced risk of cancer.

Finally, the EPA should reassess the strength of the evidence relating arsenic to elevated cancer risks in U.S. populations. This is especially important considering that recent studies have failed to confirm the expected elevated cancer risks.

# CHAPTER 7
# RECOMMENDATIONS TO UTILITIES

This study addresses several issues vital to drinking water utilities and the industry. Although this study had no direct or immediate implications for drinking water treatment, its findings can help guide the water industry in evaluating the scientific data on arsenic health effects. The information can also be useful to customers concerned about the safety of their drinking water.

Despite the fact that recent U.S. studies have failed to detect the expected arsenic-related cancer risks, there are indications that EPA intends to further lower the arsenic MCL below 10 μg/L. As a first step, the EPA plans to increase the cancer slope factor for arsenic in its model. This will justify lowering the arsenic MCL. Several states, including California and New Jersey, have already begun the process of lowering their arsenic MCL to below 10 μg/L. In fact, California is considering a standard in the parts per trillion range.

The study presented in this report provides additional information useful for assessing arsenic-related health risks in U.S. populations and can help regulators better evaluate the need to further lower the arsenic MCL.

# REFERENCES


Abernathy, C.O., D.J. Thomas, and R.L. Calderon. 2003. Health Effects and Risk Assessment of Arsenic. *Journal of Nutrition*,133(5 Suppl 1):1536S-1538S.

Agency for Toxic Substances and Disease Registries (ATSDR). 2000. *Toxicological Profile for Arsenic. Update.* Atlanta, Georgia: Department of Health and Human Services, Public Health Service.

Axelson, O., E. Dahlgren, C.D. Jannsson, and S.O. Rhenlund. 1978. Arsenic Exposure and Mortality: Case-Referent Study From a Swedish Copper Smelter. *British Journal of Industrial Medicine,* 35(1):8-15.

Bates M.N., O. Rey, M.L. Biggs, C. Hopenhayn, L.F. Moore, D. Kalman, C. Steinmous, and A. Smith. 2003. Arsenic in Drinking Water and Bladder Cancer in Argentina: Results of a Case-Control Study (Abstract). *Epidemiology*. 14(5):S123.

Bates, M.N., A.H. Smith, and K.P. Cantor. 1995. Case-Control Study of Bladder Cancer and Arsenic in Drinking Water. *American Journal of Epidemiology*, 1141(6):523-530.

Bates, M.N., A.H. Smith, and C. Hopenhayn-Rich. 1992. Arsenic Ingestion and Internal Cancers: A Review. *American Journal of Epidemiology*, 135(5):462-476.

Blejer, H.P., and W. Wagner. 1976. Case Study 4: Inorganic Arsenic – Ambient Level Approach to the Control of Carcinogenic Exposures. *Annals of the New York Academy of Sciences*, 271:179-186.

Blot, W.J., and J.F. Fraumeni. 1975. Arsenical Air Pollution and Lung Cancer. *Lancet,* 2(7926):142-144.

Borgono, J.M., P. Vicent, H. Venturina, and A. Infante. 1977. Arsenic in the Drinking Water in the City of Antofagasta: Epidemiological and Clinical Study Before and After the Installation of a Treatment Plant. *Environmental Health Perspectives*, 19(Aug.):103-109.

Breslow, N.E. 1984. Extra-Poisson Variation in Log-Linear Models. *Applied Statistics,* 33:38-44.

Braun, W. 1958. Carcinoma of the Skin and Internal Organs Caused by Arsenic; Contribution to Late Occupational Sequelae Caused by Arsenic. *Dtsch Med Wochansche,* 83(20):870-872.

Brown, K.G., and G.L. Ross. 2002. Arsenic, Drinking Water, and Health: A Position Paper of the American Council on Science and Health. *Regulatory Toxicology and Pharmacology,* 36(2):162-174.

Brown, L.M., L.M. Pottern, and W.J. Blot. 1984. Lung Cancer in Relation to Environmental Pollutants Emitted From Industrial Sources. *Environmental Research,* 34(2):250-261.

Buchet, J.P., and D. Lison. 1998. Mortality by Cancer in Groups of the Belgian Population With a Moderately Increased Intake of Arsenic. *International Archives of Occupational and Environmental Health,* 71(2):125-130.

Calabrese, E.J. 1983. Role of Epidemiologic Studies in Deriving Drinking Water Standards for Metals. *Environmental Health Perspectives*, 52(Oct.):99-106.

Centers for Disease Control 2000. EpiInfo 2000 (software). Available Online: www.cdc.gov/epiinfo/.

Chen, C.J., C.W. Chen, M.M. Wu, and T.L. Kuo. 1992. Cancer Potential in Liver, Lung, Bladder and Kidney Due to Ingested Inorganic Arsenic in Drinking Water. *British Journal of Cancer*, 66(5):888-892.

Cheng, M.L., H.Y. Ho, Y.W. Huang, F.J. Lu, and D.T.-Y. Chiu. 2003. Humic Acid Induces Oxidative DNA Damage, Growth Retardation and Apoptosis in Human Primary Fibroblasts. *Experimental Biology and Medicine*, 228(4):413-423.

Chiang, H.S., H.R. Guo, C.L. Hong, S.M. Lin, and E.F. Lee. 1993. The Incidence of Bladder Cancer in the Black Foot Disease Endemic Area in Taiwan. *British Journal of Urology*, 71(3):274-278.

Chiou, H.Y., S.T. Chiou, Y.H. Hsu, Y.L. Chou, C.H. Tseng, M.L. Wei, and C.J. Chen. 2001. Incidence of Transitional Cell Carcinoma and Arsenic in Drinking Water: A Follow-up Study of 8,102 Residents in an Arseniasis-Endemic Area in Northeastern Taiwan. *American Journal of Epidemiology*, 153(5): 411-18.

Clayton, D.G., and J. Kaldor. 1987. Empirical Bayes Estimates of Age-Standardized Relative Risks for Use in Disease Mapping. *Biometrics*, 43(3):671-81.

Cordier, S., G. Theriault, and H. Iturra. 1983. Mortality Patterns in a Population Living Near a Copper Smelter. *Environmental Research,* 31(2):311-22.

Dorn H.F., Horn J.L. 1941. The Reliability of Certificates of Death From Cancer. *American Journal of Hygiene*. 24:12-23.

Engel, R.R., and A.H. Smith. 1994. Arsenic in Drinking Water and Mortality From Vascular Disease: An Ecologic Analysis in 30 Counties in the United States. *Archives of Environmental Health,* 49(5):418-27.

Enterline, P.E., and G.M. Marsh. 1982. Cancer Among Workers Exposed to Arsenic and Other Substances in a Copper Smelter. *American Journal of Epidemiology*, 116(6):893-911.

Enterline, P., and W. Stewart. 1956. Geographic Patterns in Deaths from Coronary Heart Disease. *Public Health Reports*, 71(9):849-855.

EPA (Environmental Protection Agency). 2000. *Arsenic Occurrence in Public Drinking Water Supplies*. EPA-815-R-00-023. Washington, D.C.: U.S. Environmental Protection Agency.

Ferreccio, C., C. Gonzalez, V. Milosavjlevic, G. Marshall, A.M. Sancha, and A.H. Smith. 2000. Lung Cancer and Arsenic Concentrations in Drinking Water in Chile. *Epidemiology,* 11(6):673-679.

Frost, F., L. Harter, S. Milham, R. Royce, A.H. Smith, J. Hartley, and P. Enterline. 1987. Lung Cancer Among Women Residing Close to an Arsenic Emitting Copper Smelter. *Archives of Environmental Health*, 42(2):148-152.

Frost, F.J., T. Muller, H.V. Petersen, B. Thomson, and K. Tollestrup. 2003. Identifying US Populations for the Study of Health Effects Related to Drinking Water Arsenic. *Journal of Exposure Analysis and Environmental Epidemiology*, 13(3):213-239.

Greaves, W.W., W.N. Rom, J.L. Lyon, G. Varley, D.D. Wright, and G. Chiu. 1981. Relationship Between Lung Cancer and Distance of Residence from Nonferrous Smelter Stack Effluent. *American Journal of Industrial Medicine,* 2(1):15-23.

Greenland, S., and J. Robins. 1994a. Invited Commentary: Ecologic Studies--Biases, Misconceptions, and Counter Examples. *American Journal of Epidemiology*, 139(8):747-760.

Greenland, S., and J. Robins. 1994b. Accepting the Limits of Ecologic Studies: Drs. Greenland and Robins Reply to Drs. Piantedosi and Cohen. *American Journal of Epidemiology*, 139(8):769-771.

Guo, H.G., H.S. Chiang, H. Hu, S.R. Lptsitz, and R.R. Monson. 1997. Arsenic in Drinking Water and Incidence of Urinary Cancers. *Epidemiology,* 8(5):545-550.

Guo, H.R., and Y.C. Tseng. 2000. Arsenic in Drinking Water and Bladder Cancer: Comparison Between Studies Based on Cancer Registry and Death Certificates. *Environmental Geochemistry and Health,* 22(2): 83-91.

Haque, R., D.N. Mazumder, S. Samanta, N. Ghosh, D. Kalman, M.M. Smith, S. Mitra, A. Santra, S. Lahiri, S. Das, B.K. De, and A.H. Smith. 2003. Arsenic in Drinking Water and Skin Lesions: Dose-response Data from West Bengal, India. *Epidemiology*, 14(2):174-82.

Harper, M., and G.P. Miranda. 1990. Management of Health Risks in the Arsenic Production Industry: Modern Production in Chile in the Context of Past Experience in Britain. *Annals of Occupational Hygiene,* 34(5):471-82.

Harter, G.H., and A.M. Novitch. 1967. An Evaluation of Gay's Solution in the Treatment of Asthma. *Journal of Allergy*, 40(6):327-336.

Hennekens, C.H., and J.E. Buring. 1987. Analysis of Epidemiologic Studies: Evaluating the Role of Chance. In: *Epidemiology in Medicine* Hennekens, C.H., Buring, J.E. (Eds.), Little, Brown and Co., Boston, pp. 243-271.

Higgins, I., K. Welch, M. Oh, G. Bond, and P. Hurwitz. 1981. Influence of Arsenic Exposure and Smoking on Lung Cancer Among Smelter Workers: Pilot Study. *American Journal of Industrial Medicine,* 2(1):33-41.

Hill, B., and E.L. Fanning. 1948. Studies in the Incidence of Cancer in a Factory Handling Inorganic Compounds of Arsenic. *British Journal of Industrial Medicine*, 5:6.

Hindmarsh, J.T., and R.F. McCurdy. 1986. Clinical and Environmental Aspects of Arsenic Toxicity. *Critical Reviews in Clinical and Laboratory Science*, 23(4):315-347.

Hopenhayn-Rich, C., M.L. Biggs, A. Fuchs, R. Bergolglio, E.E. Tello, Nicolli, and A.H. Smith. 1996. Bladder Cancer Mortality Associated With Arsenic in Drinking Water in Argentina. *Epidemiology,* 7(2):117-124.

Hutchinson, J. 1888. On Some Examples of Arsenic Keratosis of the Skin and of Arsenic Cancer. *Trans Pathol Soc London,* 39:352-363.

Karagas, M.R., T.A. Stukel, J.S. Morris, T.D. Tosteson, J.E. Weiss, S.K. Spencer, and E.R. Greenberg. 2001. Skin Cancer Risk in Relation to Toenail Arsenic Concentrations in a US Population-Based Case-Control Study. *American Journal of Epidemiology*, 153(6):559-565.

Kliewer, E.V. 1992. Influence of Migrants on Regional Variations of Stomach and Colon Cancer Mortality in the Western United States. *International Journal of Epidemiology,* 21(3):442-449.

Kobayashi, J. 1957. On Geographic Relationships Between the Chemical Nature of River Water and Death Rates From Apoplexy. *Ber Ohara Inst Landw Biol.,*11:12-21.

Kurttio, P., E. Pakkala, H. Kahelin, A. Auvinen, and J. Pekkanen. 1999. Arsenic Concentrations in Well Water and Risk of Bladder and Kidney Cancer in Finland. *Environmental Health Perspectives,* 107(9):705-710.

Lamm S.H., D.M. Byrd, M.B. Kruse, M. Feinleib, and S.H. Lai. 2003. Bladder Cancer and Arsenic Exposure: Differences in the Two Populations Enrolled in a Study in Southwest Taiwan. *Biomedical and Environmental Science*. 16:355-368.

Langford, I.H., A.H. Leyland, J. Rasbash, and H. Goldstein. 1999. Multi-level Modelling of the Geographical Distributions of Disease. *Applied Statistics*, 48 (2): 253-268.

Lawson, A., W. Browne, and C. Rodeiro. 2003. Disease Mapping With WinBUGS and MLwiN. John Wiley & Sons.

Lee, A.M., and J.F. Fraumeni. 1969. Arsenic and Respiratory Cancer in Man--An Occupations Study. *Journal of the National Cancer Institute,* 42(6):1045-1052.

Lewis, D.R., and J.W. Southwick, Ouellet-Hellstrom R, Rench J, Calderon RL. 1999. Drinking Water Arsenic in Utah: a Cohort Mortality Study. *Environmental Health Perspectives*, 107(5):359-365.

Liu, P., and Z.C. Han. 2003. Treatment of Acute Promyelocytic Leukemia and Other Hematologic Malignancies With Arsenic Trioxide: Review of Clinical and Basic Studies. *International Journal of Hematology*, 78(1):32-39.

Long, L.H. 1988. Migration and Residential Mobility in the United States. New York: Russell Sage Foundation (Census Monograph Series: The Population of the United States in the 1980's).

Lu, F.J. 1990. Blackfoot Disease: Arsenic or Humic Acid? *Lancet*, 336(8707):115-116.

Lyon, J.L., J.L. Fillmore, and M.R. Klauber. 1977. Arsenicical Air Pollution and Lung Cancer, *Lancet,* 2(8043):869.

Mabuchi, K., A.M. Lilienfeld, and L.M. Snell. 1979. Lung Cancer Among Pesticide Workers Exposed to Inorganic Arsenicals. *Archives of Environmental Health*, 34:312-320.

Marx, A., and R.R. Neutra RR. 1997. Magnesium in Drinking Water and Ischemic Heart Disease. *Epidemiology Reviews,* 19(2):258-272.

Matanoski, G., E. Landau, J. Tonascia, C. Lazxar, E. Elliot, and W. McEnroe. 1981. Cancer Mortality in an Industrial Area of Baltimore. *Environmental Research*, 25(1):8-28.

Milham, S. 1978. Industrial Toxins and the Community: Arsenic Contamination Near a Copper Smelter. In *Occupational Safety and Health Symposia 1977*. Publication 78-164. U.S. Department of Health and Human Services.

Milham, S. 1983. *Review and Analysis of Literature on Health of Populations Living Near Arsenic Emitting Sources*. Seattle, Wash.: Washington State Department of Health.

Moore, L.E., M. Lu, and A.H. Smith. 2002. Childhood Cancer Incidence and Arsenic Exposure in Drinking Water in Nevada. *Archives of Environmental Health*, 57(3):201-206.

Morales, K.H., L. Ryan, T.L. Kuo, M.M. Wu, and C.J. Chen. 2000. Risk of Internal Cancers From Arsenic in Drinking Water. *Environmental Health Perspectives*, 108:655-661.

Moriyama I.M., W.S. Baum, W.M. Haenszel, and B.F. Mattison. 1958. Inquiry Into Diagnostic Evidence Supporting Medical Certifications of Death. *American Journal of Public Health.* 48(10):1376-1387.

Morton, W., G. Starr G, D. Pohl, J. Stoner, S. Wagner, P. Weswig. 1976. Skin Cancer and Water Arsenic in Lane County, Oregon. *Cancer,* 37(5):2523-2532.

Mushak, M. 2000. *Arsenic and Old Laws. A Scientific and Public Health Analysis of Arsenic Occurrence in Drinking Water, it's Health Effects, and EPA's Outdated Arsenic Tap Water Standard.* New York, NY: National Resources Defense Council.

Newman, J.A., V.E. Archer, G. Saccomanno, M. Kuschner, O. Auerbach, R.T. Grondahl, and JC Wilson. 1976. Histologic Types of Bronchogenic Carcinoma Among Members of Copper-Mining and Smelting Communities. *Annals of the NY Academy of Sciences*, 271:260-8.

Neubauer, O. 1947. Arsenical Cancer: A Review. *British Journal of Cancer*, 1:192-251.

NRC (National Research Council). 1977. *Drinking Water and Health, Volume 1*. Washington, DC: National Academy Press.

NRC (National Research Council). 1999. *Arsenic in Drinking Water*. Washington, D.C.: National Academy Press.

NRC (National Research Council). 2001. *Arsenic in Drinking Water: 2001 Update.* Washington, D.C.: National Academy Press.

Mushak, M. 2000. Arsenic and Old Laws. A Scientific and Public Health Analysis of Arsenic Occurrence in Drinking Water, its Health effects, and EPA's Outdated Arsenic Tap Water Standard. New York, NY: National Resources Defense Council.

Ott, M.G., B.B. Holder, and H.L. Gordon. 1974. Respiratory Cancer and Occupations Exposure to Arsenicals. *Archives of Environmental Health*, 29:250-255.

Percy, C., E. Stanek, and L. Gloeckler. 1981. Accuracy of Cancer Death Certificates and Its Effect on Cancer Mortality Statistics. *American Journal of Public H*ealth, 71(3):242-250.

Pershagen, G., C.G. Elinder, and A.M. Bolander. 1977. Mortality in a Region Surrounding an Arsenic Emitting Plant. *Environmental Health Perspectives,* 19(Aug.):133-137.

Pinto, S.S., P.E. Enterline, V. Henderson, and M.O. Varner. 1977. Mortality in Relation to a Measured Arsenic Trioxide Exposure. *Environmental Health Perspectives*, 19(Aug.):127-130.

Pinto, S.S., V. Henderson, and P.E. Enterline. 1978. Mortality Experience of Arsenic Exposed Workers. *Archives of Environmental Health*, 33(3):325-331.

Pocock, S.J., D.G. Cook, and S.A.A. Beresford. 1981. Regression of Area Mortality Rates on Explanatory Variables: What Weighting is Appropriate? *J R Stat Soc Ser C Applied Statistics*, 30(3):286-295.

Polissar, L. 1980. The Effect of Migration on Comparison of Disease Rates in Geographic Studies in the United States. *American Journal of Epidemiology,* 111(2):175-182.

Polissar, L., K. Lowry-Coble, D.A. Kalman, J.P. Hughes, G. van Belle, D.S. Covert, T.M. Burbacher, D. Bolgiano, and N.K. Mottet. 1990. Pathways of Human Exposure to Arsenic in a Community Surrounding a Copper Smelter. *Environmental Research*, 53(1):29-47.

Polissar, L., R.K. Severson, and Y.T. Lee. 1978. *Cancer Incidence in Relation to a Smelter.* Dept. of Biostatistics, School of Public Health and Community Medicine. University of Washington, Technical Report No. 28.

Rochelle, P.A., A. A. Mofidi, M. M. Marshall, S. J. Upton, B. Montelone, K. Woods, and G. Di Giovanni. 2004. *An Investigation of UV Disinfection and Repair in* Cryptosporidium parvum. Denver, CO.: Awwa Research Foundation.

Rom, W.N., G. Varley, J.I. Lyon, and S. Shopkow. 1982. Lung Cancer Mortality Among Residents Living Near the El Paso Smelter. *British Journal of Industrial Medicine*, 39(3):269-272.

Rosenberg, H. 1989. Improving Cause-of-Death Statistics. *American Journal of Public Health*, 79(5):563-564.

Roth, F. 1957. After-Effects of Chronic Arsenicism in Moselle Vintners. *Dtsch Med Wochenischr.,* 82(6):211-217.

Schachter, J.P., and J.J. Kuenzi. 2002. Seasonality of Moves and the Duration and Tenure of Residence: 1996. Working Paper No. 69, U.S. Bureau of the Census, Washington, D.C.

Steinmaus, C., Y. Yuan, M.N. Bates, and A.H. Smith. 2003. Case-Control Study of Bladder Cancer and Drinking Water Arsenic in the Western United States. *American Journal of Epidemiology*, 158(12):1193-1201.

Tokudome, S. and M. Kuratsune. 1976. A Cohort Study on Mortality from Cancer and Other Causes Among Workers at a Metal Refinery. *International Journal of Cancer*, 17(3):310-317.

Tollestrup, K., J.R. Daling JR, and J. Allard J. 1995. Mortality in a Cohort of Orchard Workers Exposed to Lead Arsenate Pesticide Spray. *Archives of Environmental Health*, 50(3):221-229.

Tollestrup, K., F.J. Frost, L.C. Harter, and G. McMillan. Mortality in Children Residing Near the ASARCO Copper Smelter in Ruston, Washington. *Archives of Environmental Health* (in press).

Tseng, W.P., H.M. Chu, S.W. How, J.M. Fong, and S. Yeh. 1968. Prevalence of Skin Cancer in an Endemic Area of Chronic Arsenicism in Taiwan. *Journal of the National Cancer Institute,* 40(3):453-463.

Uitenbroek, D. G. 1997. "SISA (Simple Interactive Statistical Analysis)-Binomial," Available Online: http://home.clara.net/sisa/sampshlp.htm

U.S. Department of Commerce, Bureau of the Census. *USA Counties 1998*. Washington D.C.: U.S. Department of Commerce, Economics and Statistics Administration.

U.S. Department of Commerce, Bureau of the Census. *County and City Data Book [United States] 2000*. Washington D.C.: U.S. Department of Commerce, Economics and Statistics Administration.

U.S. Department of Commerce, Bureau of the Census. *County and City Data Book [United States] 1956, 1967, 1972, 1977, 1983* (computer files). Washington D.C.: U.S. Department of Commerce, Economics and Statistics Administration.

U.S. Department of Commerce, Bureau of the Census. 1989. *County and City Data Book [United States] 1988 [computer file]*. Washington D.C.: U.S. Department of Commerce, Economics and Statistics Administration.

U.S. Department of Commerce, Bureau of the Census. 1990. Census of Population and Housing, "1990 Population and Housing Unit Counts: United States", (CPH-2). Washington D.C.

U.S. Department of Commerce, Bureau of the Census. Table C3. Per Capita Income by County, Income in 1989 CPI-U Adjusted Dollars. *Censuses of Population and Housing, 1960, 1970, 1980, 1990*. Washington D.C.: U.S. Department of Commerce, Economics and Statistics Administration.

# ABBREVIATIONS

| | |
|---|---|
| AOED | Arsenic Occurrence and Exposure Database |
| AWWA | American Water Works Association |
| AwwaRF | Awwa Research Foundation |
| BCC | basal cell carcinoma |
| BFD | blackfoot disease |
| EPA | U.S. Environmental Protection Agency |
| L | liter |
| MCL | maximum contaminant level |
| MCMC | Markov Chain Monte Carlo methods |
| μ | microgram |
| μg/L | micrograms per liter |
| NA | not available |
| NCI | National Cancer Institute |
| NRC | National Research Council |
| NRDC | National Resources Defense Council |
| PMSA | primary metropolitan statistical area |
| RIGLS | restrictive iterative generalized least squares |
| SCC | squamous cell carcinoma |
| SDWIS | Safe Drinking Water Information System |
| SEER | Surveillance, Epidemiology and End Results |
| SIR | standardized incidence ratio |
| SMR | standardized mortality ratio |
| U.S. | United States |

IWA Publishing's authorised EU representative for General Product Safety Regulations is Diane D'Arras, 15 rue Duret, 75116 Paris, France, e-mail: safety@iwap.co.uk.

Printed and bound by CPI Group (UK) Ltd, Croydon, CR0 4YY
17/09/2026
02218808-0009